T0271008

Quality Assessment and Security in Industrial Internet of Things

This book highlights authentication and trust evaluation models in the Industrial Internet of Things. It further discusses data breaches and security issues in various Artificial Intelligence–enabled systems and uses Blockchain to resolve the challenges faced by the Industrial Internet of Things. The text showcases performance quality assessment for the Industrial Internet of Things' applications.

This book:

- Discusses and evaluates different quality assessment systems and authentication of smart devices.
- Addresses data handling, data security, confidentiality, and integrity of data in the Industrial Internet of Things.
- Focuses on developing framework and standardization of quality assessment for diverse Internet of Things-enabled devices.
- Explains the designing, developing, and framing of smart machines, that are equipped with tools for tracking and logging data to provide advanced security features.
- Presents the convergence of the Internet of Things toward Industry 4.0 through quality assessment via analyzing data security and identifying vulnerabilities.

It is primarily written for graduate students and academic researchers in the fields of electrical engineering, electronics, and communications engineering, industrial and production engineering, computer science, and engineering.

Quality Assessment and Security in Industrial Internet of Things

Edited by
Sudan Jha, Sarbagya Ratna Shakya,
Sultan Ahmad, and Zhaoxian Zhou

CRC Press
Taylor & Francis Group
Boca Raton London New York

CRC Press is an imprint of the
Taylor & Francis Group, an **informa** business

Designed cover image: metamorworks/Shutterstock

First edition published 2025
by CRC Press
2385 NW Executive Center Drive, Suite 320, Boca Raton FL 33431

and by CRC Press
4 Park Square, Milton Park, Abingdon, Oxon, OX14 4RN

CRC Press is an imprint of Taylor & Francis Group, LLC

© 2025 selection and editorial matter, Sudan Jha, Sarbagya Ratna Shakya, Sultan Ahmad and Zhaoxian Zhou; individual chapters, the contributors

ISBN: 9781032538730 (hbk)
ISBN: 9781032870359 (pbk)
ISBN: 9781003530572 (ebk)

DOI: 10.1201/9781003530572

Typeset in Sabon
by codeMantra

Contents

Note from the editors

Welcome to the groundbreaking exploration of the Industrial Internet of Things (IIoT) in our comprehensive volume titled *Quality Assessment and Security in Industrial Internet of Things*. As we all have been witnessing the profound integration of smart technology into various aspects of daily life, the significance of ensuring the reliability, security, and privacy of IoT systems becomes paramount. Our book delves into the intricate landscape of IIoT, offering insights, methodologies, and solutions to address the evolving challenges in this domain.

Our book starts with a fundamental understanding of IoT systems and their significance in revolutionizing industries such as manufacturing, healthcare, energy, transportation, and beyond. We unravel the complexities surrounding the quality assessment of smart devices, emphasizing the crucial role of parameters like Quality of Service (QoS) in ensuring optimal performance and reliability. Readers will gain valuable insights into assessing, evaluating, and enhancing the quality of smart devices, thereby contributing to the advancement of IIoT ecosystems.

One of the central themes of our book is the critical importance of cybersecurity in IIoT environments. We meticulously examine the security and privacy challenges inherent in IoT systems, offering in-depth analyses of current attacks, vulnerabilities, and intrusion detection mechanisms. Through innovative approaches and practical methodologies, we empower readers to fortify IIoT systems against emerging threats, safeguarding sensitive data, and ensuring the integrity of critical infrastructure.

Furthermore, our book ventures into the realm of artificial intelligence (AI) and its transformative potential in enhancing the security posture of IIoT networks. From intelligent intrusion detection systems to advanced threat analysis techniques, we explore the role of AI-driven solutions in bolstering the resilience of IIoT ecosystems against malicious actors and cyber threats.

As editors, we are honored to present this comprehensive compilation of research, insights, and practical solutions aimed at advancing the field of IIoT. Whether you are a seasoned professional seeking to deepen your understanding of IIoT security or a newcomer navigating the complexities of smart technology, this book serves as an indispensable guide on your journey toward mastering the intricacies of the IIoT.

We welcome our readers to join us in a transformative exploration of Quality Assessment and Security Challenges in Industrial IoT, shaping the future of connected industries and driving innovation in the digital era.

About the editors

Sudan Jha: With over 23 years of combined teaching, research, and industrial experience, Sudan Jha is a senior member of IEEE and a Professor in the Department of Computer Science & Engineering at Kathmandu University, Nepal. His previous affiliations include KIIT University, Chandigarh University, and Christ University. He was Technical Director at Nepal Television, Principal at Nepal College of IT, and Individual Consultant at Nepal Telecom Authority. Prof. Jha is dedicated to advancing higher education quality and actively works on smart platforms. His extensive research portfolio comprises 80+ SCI, SCIE-indexed research papers and book chapters in international peer-reviewed journals and conferences. He serves as a Co-Editor-in-Chief of an international journal. He also serves as Guest Editor for SCIE/ESCI/SCOPUS-indexed journals. With three patents to his name, he has authored and edited six books on cutting-edge topics in IoT, 5G, and AI, published by reputed publishers. His research has also secured funding for two international projects. Additionally, Prof. Jha contributes as a keynote speaker at more than 40 international conferences. In addition, he has also delivered faculty development programs, short-term training programs, and workshops in national and international conferences and universities. He holds certifications in Microservices Architecture, Data Science, and Foundations of Artificial Intelligence. His primary research interests encompass Quality of Services in IoT-enabled devices, Neutrosophic theory, and Neutrosophic Soft Set Systems.

Sarbagya Ratna Shakya: Sarbagya Ratna Shakya is an Assistant Professor at Eastern New Mexico University, New Mexico, USA. He received a B.Eng. in Electronics Engineering from National College of Engineering, Tribhuvan University of Nepal in 2009 and an M.Eng. in Computer Engineering from Nepal College of Information Technology, Pokhara University of Nepal in 2014. He received his Ph.D. in Computational Science (Computer Science) from the School of Computer Science and Computer Engineering, University of Southern Mississippi, USA. He has more than 10 years of teaching experience in undergraduate-level courses. His research interest includes machine learning, deep learning, image processing, Internet of Things (IoT), and Industrial IoT and has published journal papers, conference papers, and book chapters in different domains on applied machine learning and deep learning.

Sultan Ahmad: Sultan Ahmad (Member, IEEE) received a Ph.D. degree in CSE from Glocal University and a Master of Computer Science and Applications degree (Hons.) from Aligarh Muslim University, India. He is currently a Faculty Member of the Department of Computer Science, College of Computer Engineering and Sciences, Prince Sattam Bin Abdulaziz University, Al-Kharj, Saudi Arabia. He is also an Adjunct Professor at Chandigarh University, Gharuan, Punjab, India. He has more than 15 years of teaching and research experience. He has around 105 accepted and published research papers and book chapters in reputed SCI-, SCIE-, ESCI-, and SCOPUS-indexed journals and conferences. He has an Australian patent and a Chinese patent in his name also. He has authored and edited five books that are available on Amazon. He has presented his research papers at many national and international conferences. His research interests include intelligence computing, big data, machine learning, and Internet of Things. He is a member of IACSIT and the Computer Society of India.

Zhaoxian Zhou: Dr. Zhaoxian Zhou is a dedicated scholar and educator in the field of Electrical and Computer Engineering. With a comprehensive academic background spanning renowned institutions worldwide, Dr. Zhou brings a wealth of expertise and experience to his role. He graduated with a Bachelor of Engineering in Electrical Engineering from the University of Science and Technology of China in 1991, a Master of Engineering in Electrical Engineering from the National University of Singapore in 1999, and a Ph.D. in Electrical and Computer Engineering from the University of New Mexico in 2005. Dr. Zhou's professional journey transitioned from industry to academia. Started his career as an electrical engineer at the China Research Institute of Radiowave Propagation in 1991, he joined the faculty at the University of Southern Mississippi in 2005. Currently, he serves as a Full Professor in the School of Computing Sciences and Computer Engineering. His research interests encompass a diverse range of topics, including wireless sensor networks, human activity recognition, face recognition, image processing, big data analytics, machine learning, high-performance computing, wearable devices, electromagnetics, and engineering education. Dr. Zhou's scholarly endeavors have resulted in more than 50 publications in esteemed academic journals and conference proceedings. Dr. Zhou is deeply committed to educating the next generation of engineers. His teaching portfolio covers a broad spectrum of subjects, including introduction to computer engineering, circuit analysis, digital electronics, analog and digital communications, Internet of Things, digital signal processing, power electronics, alternative energy, microwave engineering, electromagnetic fields, analytical methods, and statistical techniques. Dr. Zhou's teaching philosophy emphasizes mentorship, collaboration, and inclusive and active learning, inspiring students to excel in their academic endeavors. Dr. Zhou's contributions to the engineering community have been recognized through various awards and honors, including the IEEE APS RMTG Award, the SUMMA Graduate Fellowship in Advanced Electromagnetics, and the Outstanding New Teacher Award from ASEE-SE, Faculty Excellence Award from USM, and multiple teaching grants. He actively engages in scholarly activities as a reviewer for leading publishers, journals, and conferences, contributing to the advancement of his field. As a senior member of the Institute of Electrical and Electronics Engineers (IEEE), a member of the HKN Honor Society of Electrical and Computer Engineering, and a program evaluator for ABET Inc., Dr. Zhou exemplifies the values of academic excellence, integrity, and service. His dedication to fostering learning and innovation extends beyond the classroom, as he continues to mentor aspiring engineers and contribute to the scholarly community.

Contributors

Mohammad Mazhar Afzal
School of Computer Science, Glocal
  University
Saharanpur, UP, India

Sultan Ahmad
Department of Computer
  Science
College of Computer Engineering
  and Sciences
Prince Sattam bin Abdulaziz
  University
Alkharj, Saudi Arabia
University Center for Research and
  Development (UCRD)
Department of Computer Science
  and Engineering
Chandigarh University, Gharuan,
  Mohali, Punjab, India

Ilemona Atawodi
School of Computing Sciences and
  Computer Engineering
University of Southern Mississippi
Hattiesburg, Mississippi

Tejinderpal Singh Brar
Department of Computer
  Applications
Chandigarh Group of Colleges
Landran, Punjab, India

Jyotir Moy Chatterjee
Department of Computer Science
  and Engineering
Graphic Era University
Dehradun, Uttarakhand, India

Rachit Garg
School of Computer Science and
  Engineering
Lovely Professional University
Phagwara, Punjab, India

Bidyut Gupta
School of Computing
Southern Illinois University
Carbondale, Illinois

Md. Alimul Haque
Department of Computer Science
Veer Kunwar Singh University
Ara, Bihar, India

Shameemul Haque
Al-Hafeez College
Ara, Bihar, India

Nthatisi Magaret Hlapisi
Electronics and Electrical
  Engineering
Lovely Professional University
Phagwara, Punjab, India

Sudan Jha
Department of Computer Science
and Engineering
Kathmandu University
Kathmandu, Nepal

Kamini
Chandigarh Group of Colleges
Landran, Punjab, India

Mohammad Shuaib Khan
Chitkara University
Institute of Engineering and
Technology
Sangrur, Punjab, India

Abhishek Kumar
Department of Computer Science
Engineering
Chandigarh University
Sahibzada Ajit Singh Nagar,
Mohali, India

Badri Raj Lamichhane
Department of Electronics &
Computer Engineering
Institute of Engineering
Tribhuvan University
Kirtipur, Nepal

Gauri Mathur
Department of Computer Science
Engineering
Lovely Professional University
Phagwara, Punjab, India

Saydul Akbar Murad
School of Computing Sciences and
Computer Engineering
University of Southern Mississippi
Hattiesburg, Mississippi

Shabdapurush Poudel
Kathmandu University
Dhulikhel, Nepal

Deepak Prashar
School of Computer Science and
Engineering
Lovely Professional University
Phagwara, Punjab, India

Nick Rahimi
School of Computing Sciences and
Computer Engineering
University of Southern Mississippi
Hattiesburg, Mississippi

Manik Rakhra
Department of Computer Science
Engineering
Lovely Professional University
Phagwara, Punjab, India

Rabie A. Ramadan
Department of Information Systems
Cairo University
Cairo, Egypt

Pramod Rathore
Department of Computer Science
Engineering
Manipal University
Jaipur, Rajasthan, India

Indranil Roy
Department of Computer Science
Southeast Missouri State University
Capte Girardeun, Missouri

Nidhi Sagarwal
Department of Forensic Sciences
Lovely Professional University
Phagwara, Punjab, India

Sarbagya Ratna Shakya
Department of Mathematical
Sciences
Eastern New Mexico University
Portales, New Mexico

Raj Karan Singh
Department of Computer Science
  Engineering
Lovely Professional University
Phagwara, Punjab, India

Anil Kumar Sinha
Department of Computer Science
Veer Kunwar Singh University
Ara, Bihar, India

Deepa Sonal
Department of Computer Science
Patna Women's College
Patna, Bihar, India

P. Srinivas Kumar
Department of Computer Science
  and Engineering
Amrita Sai institute of science and
  technology
Vijayawada, Andhra Pradesh, India

R. Sujatha
School of Computer Science
  Engineering and Information
  Systems
Vellore Institute of Technology
Katpadi, Vellore, India

Shipra Shivkumar Yadav
Department of Computer Science
Lincoln University
Kota Bharu, Malaysia

Zhaoxian Zhou
School of Computing Sciences and
  Computer Engineering
University of Southern Mississippi
Hattiesburg, Mississippi

Introduction to Industrial Internet of Things (IIoT) – toward the future internet

Sarbagya Ratna Shakya and Sudan Jha

1.1 INTRODUCTION

The Internet of Things (IoT), which was first conceptualized in 1999 [1], has a crucial application in the industry. The Industrial Internet of Things (IIoT) refers to the interrelated, automated use of sensors, devices, and machines that run on industrial applications to increase efficiency and reliability of manufacturing. It provides connectivity between systems, machines, and people in companies. Many companies need to improve asset management and maintenance by reducing waste and cost to improve their competitive advantage. IIoT helps to increase operational efficiency in the manufacturing industry by integrating information technology with operational technology [2]. IIoT is also commonly known as Industry 4.0, which works on interconnected smart machinery, embedding sensing devices, data analytics, and automated decision-making with edge computing technology to increase productivity, make supply chain efficient, and improve distribution, capacity, resource management [3], worker safety, and return on investment [4]. IIoT integrates technology related to Artificial Intelligence (AI) with manufacturing learning and responding, analyzing big data in operations and productions [5]. This integration improves manufacturing performance, scalability, and ideas behind evolving and improving the business by capturing data from sensors and communicating accurately and consistently [6]. The raw data collected from these sensors and assets consist of the digital footprints of the manufacturing product and process. These data, when analyzed, can provide information regarding the production and supply chain that can be implemented in decision-making and optimization. Although IIoT has added value to business operations, the adoption of these digital technologies has still been difficult for businesses. In emerging economies, the slow and low implementation has slowed the progress in these fields [7]. Also, interconnecting different sensors, actuators, and controllers with the production lines and equipment to perform the manufacturing process automatically without any or minimum human intervention has added challenges in terms of safety and security of industrial

DOI: 10.1201/9781003530572-1

1

production and automation [8]. Because of this reason, some companies have shown resistance and reservations in adopting these technologies. However, with the improvement of these technologies, many manufacturers are now switching to implementing IIoT in their manufacturing process.

Recent developments in different sectors such as process modeling, cloud computing, cloud data, data analysis, supercomputing, AI including deep learning, and communication like 5G/6G have made IIoT possible to optimize industry processes and services. Interaction between machine-to-machine (M2M), communication, and data sharing has been possible. These collected data from sensors are then sent to a centralized cloud-based system or database and are analyzed by a cloud-hosted application that provides key information to monitor machines or analyze key performance of the automated system.

1.2 IIOT ARCHITECTURE AND COMPONENTS

IIoT technology consists of four different components:

1. Device layers: These consist of edge devices involved in industrial operations.
2. Network layers: These consist of different communication mediums such as Wi-Fi, Bluetooth, and transmission mediums.
3. Processing layers: In these, data are processed on different platforms such as the cloud, data centers, and web services.
4. Application layers: In these, the technology is applied in different applications like smart cities and smart manufacturing.

Figure 1.1 illustrates the different layers of IIoT architecture and their components. The details about these different layers of IIoT are described below.

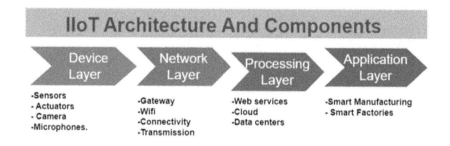

Figure 1.1 IIoT architecture and components.

Device Layers: These layers consist of physical devices involved in the industrial sectors such as sensors, actuators, cameras, processors, meters, microphones, controllers, and edge devices that collect data and communicate those data in the network through edge or IIoT gateways. These layers are located close to the edge of the system. Most of the smart devices used in this layer communicate wirelessly with built-in networking capabilities, whereas some legacy devices that do not have these wireless network capabilities usually connect through serial interfaces. These devices sense or control things at the edge in the real world and communicate wirelessly to the networks. These data can be collected from different objects or environments.

Network Layers: The data collected from the sensors are aggregated and converted into digital streams and are transmitted over the internet in the network layer. These layers include many security-related tools and applications such as authentication, encryption, and malware protection to ensure the security of the data. These layers also require high processing power due to the large amount of processed data collected from various sensors which are streamed continuously and then transferred through different network devices. The network layers should have gateways that are portable, flexible, and capable of supporting various platforms, and should be able to adapt to different environmental conditions. Some analytics and data management services are also included in these layers for analyzing the input data streams in real time. Some edge computing applications can preprocess and perform analytics on the received data with limited processing capability.

Processing Layers: Once the data are collected, they are transferred through the network layer to the cloud domain for deeper and more comprehensive processing. These high-quality data are stored, processed, analyzed, managed, and drawn conclusions using AI. These layers also include cloud-based technologies and web services based on on-premise data centers or hybrid systems. Different platforms, such as cloud servers and databases, are used to integrate devices and data across the IIoT system. In these layers, the data are processed and analyzed, and useful insights are extracted from those data and then applied to business applications, integrating automated processes.

Application Layers: These layers consist of the user-facing components. These layers provide the interface to view the conclusions from the analyzed data in the form of reports, analyses, and statistics that are interpreted and implemented in different applications [9]. Application layers comprise the front-end enterprise applications. Applications can access data from processing layers in real time, providing the industry smart manufacturing process with smart components and equipment. These layers may also have central data repository with data processing and analytics capability. The end-use capabilities of the IIoT such as predictive maintenance, remote monitoring, energy management, asset performance, and deeper performance insights are illustrated in these layers.

1.3 KEY TECHNOLOGIES IN IIOT

The ecosystem of IIoT consists of different technologies such as automation, internet of services, wireless technology, computing, cyber-physical systems (CPS), IoT, and cloud computing. Among these, the development and advancement of CPS, edge computing, and cloud computing play a bigger role in the development of IIoT to achieve increased efficiency in the manufacturing process [10]. Figure 1.2 shows some of the related key technologies related to IIoT, and the following are the details of some of the emerging technologies related to IIoT.

A. *Cyber-Physical Systems (CPS):* CPS is a smart system featuring electronics and embedded software interacting with physical components that are connected with or without the internet. It offers real-time data collection, analysis, and transparency across every aspect of the manufacturing operations. CPS is used in industrial sectors for command and control of the mechanism, security, resiliency, and automation of the system [11]. It provides critical elements such as computing and networking for IIoT implementation. This system consists of

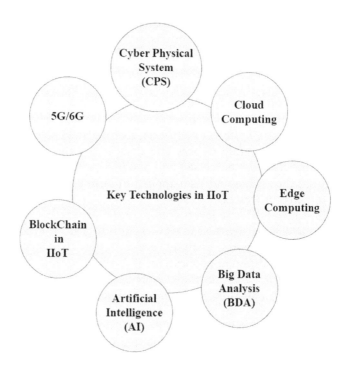

Figure 1.2 Key technologies in IIoT.

two parts. The first part is the cyber system which includes control, networking, and computing networks that are capable of operating, interconnection, and intelligence of the industrial systems, and the second part is the physical system that includes manufacturing and automation systems [12]. This system collects data through sensors and saves, evaluates, and processes them in these networks. CPS helps to increase the efficacy of traditional physical plants by regulating, self-monitoring, and functioning autonomous operations. It helps to control production systems remotely through the internet-based architecture. Some examples of CPS include autonomous vehicles, smart grids, smart cities, and smart manufacturing.

B. *Cloud Computing*: Cloud computing is a model that provides infrastructure, platform, and software services having computational, networking, and storage capabilities. Cloud computing uses interconnected remote servers hosted on the internet to store, manage, and process data or information. Cloud computing has provided a greater platform and infrastructure to monitor, actuate, and control the IIoT-enabled devices and processes in real time for manufacturing and industrial automation. Cloud computing provides services through Infrastructure as a Service (IaaS), Platform as a Service (PaaS), and Software as a Service (SaaS). Some of the widely used IaaS services are Microsoft Azure, Amazon Wave Services (AWS), and Google Cloud. These are mostly used as servers for networking and storage. PaaS services are used for database management, tool development, and system operation. They are used for developing, testing, and managing software and their applications. In this setup, users can access software and hardware tools primarily used for app development provided by third parties through the internet. Microsoft Azure and AWS are the two most popular cloud computing services that provide these services. SaaS is an application delivery model used for out-of-the-box solutions. It provides access to software or app hosting and is provided to the end users through the internet. Some of the examples include Salesforce and Microsoft Office. Many companies rely on cloud computing to implement their IIoT technology on machines to run their business, as it offers benefits such as resource pooling, resource access, and elasticity [13] with improved computing power and rendering capabilities.

C. *Edge Computing*: Edge computing is an extension of cloud computing which helps in enabling faster decision-making and analyzing data at the end device level [14]. It allows to perform computing at the end device level like physical devices and at local gateway level [15]. This implementation proves beneficial in real-time applications improving performance by reducing data transmission and power consumption [16]. Edge computing has the ability to collect an adequately large amount of quality data that can be used for training AI models [17] with minimum computational time, limited processing requirements, and improved resource management. This also helps to optimize network performance by lowering bandwidth costs, reducing communication latency, and improving connectivity and privacy. Edge computing brings cloud services to the edge of the system by improving the Quality of Service (QoS) of the system. Additionally, it will reduce the processing burden and helps to save resources [18]. Edge computing is used to process data at the edge, reducing the workload on cloud computing.

D. *Big Data Analytics (BDA)*: Because of the extensive use of sensors and IoT devices in IIoT, there is a vast amount of data that need to be managed, processed, analyzed, and visualized to extract insights [19]. These raw data can be organized and processed, and then meaningful conclusions can be drawn which can give meaningful value for analyzing and decision-making. These big data can be processed in the manufacturing industry to achieve real-time implementation and production optimality with increased efficiency. In IIoT, BDA provides tools and analytic operations to visualize real-time actionable knowledge. When integrated into business operations, it facilitates decision-making aimed at maximizing profits and evolving the business [20].

E. *Artificial Intelligence (AI) and Machine Learning (ML)*: Integration of AI and ML has contributed greatly in IIoT to automate a wide range of industry activities [21]. Different ML algorithms are used to retrieve insights and derive beneficial information by analyzing the huge data collected from their sensors. This information is used to make effective and efficient decisions by developing predictive and recommendation models through training that helps in improving reliability, increasing productivity, troubleshooting issues in real time, and gaining customer satisfaction by improving real-time response.

F. *Blockchain in IIoT*: As security has been one of the main challenges faced in the implementation of IIoT, much research has been carried out to implement blockchain technology to make the processing chain more secure, traceable, and transparent. Blockchain technology provides encrypted data sharing with privacy preservation. Different schemes have been implemented in IoT applications, such as PTAS [22], which uses blockchain-based public-key infrastructure (PKI); secureSVM [23], which uses a public-key cryptosystem; BDKMA [24], a blockchain-based distributed key management system based on the idea of authorization mode and group access pattern; and TSMPC [25], a threshold secure multi-party computing protocol with secret sharing. Different other IoT-based applications use other schemes that use attributes-based access control schemes, reputation, and trust mechanisms for blockchain. These schemes can also be implemented in IIoT to ensure security in data transfer. Different blockchain-based IIoT architectures have been proposed to address the interoperability, integrity, security, and privacy of the data [26,27].

G. *5G/6G for IIoT*: Wireless communication has been one of the important factors for seamless, pervasive, and scalable connectivity among different sensors, machines, and other devices for the application of IIoT in smart factories and Industry 4.0. 5G provides a highly reliable, secure, and high-speed communication network that allows IIoT devices to collect, analyze, and communicate data from sensors and devices. It provides significant reliable communications with low latency, support for Ethernet integration, time-sensitive networking (TSN), and security. These networks provide high bandwidth to connect and control different IIoT devices. Several initiatives [28] have been developed for the integration of 5G technologies in manufacturing industries. 5G can provide ultra-low latency and high data rate required for IIoT [29]. Adopting 5G will benefit industries by optimizing the production process, gaining insights from the collected data, and increasing its bottom line.

1.4 KEY REQUIREMENTS OF IIOT

Integration of IIoT in the industry has many challenges that need to be addressed to make the system efficient enough to communicate with each other. Challenges such as security, privacy, and reliability should be addressed to integrate these in highly growing heterogeneous technologies in the system in real time. Also, implementing IIoT requires a huge capital investment in systems that can interact with the industry's physical equipment that helps in decision-making. Here are some key requirements and factors that should be considered for successful adoption and implementation of IIoT in the industry.

A. *Interoperability*: Implementing IIoT requires interoperability that can connect and transfer data between different devices and systems, manufactured by various vendors. These systems should support integration of various assets and complex systems with different protocols and work reliably with the systems.

B. *Scalability*: IIoT implementation requires new components and devices. More sensors and devices will be integrated into the system, providing vast amounts of data. The system should be capable of handling this huge data efficiently without any trouble. Additionally, it should be able to scale the IIoT system by integrating the existing system with necessary updates and expanded processes.

C. *Security*: One of the key requirements and challenges faced by IIoT is security. Safeguarding connected devices and networks and transferring data are key for adopting IIoT in manufacturing industries. IIoT security requirements include security in terms of network, models and methodologies, data transport, data flow control, data privacy, maintainability, access control, and authentication [30]. For these security goals and requirements, models such as the CIA Triad [31], which include confidentiality, integrity, and availability of data, are being considered. Confidentiality involves encrypting the data for transfer, controlling access, and ensuring network privacy and isolation. Integrity includes consistency, authenticity, and accuracy. Availability encompasses the overall trustworthiness, redundancy, and decentralization of the system.

1.5 ADVANTAGES OF IIOT

IIoT provides different benefits in terms of scalability, connectivity, reliability, time-saving, cost, security, and safety.

A. *Predictive Maintenance*: Implementation of IIoT can provide information regarding the assets and machinery from the data gathered, which can be used to predict potential equipment failure. Equipment data are used and analyzed to detect and predict the potential abnormalities and critical issues that may arise using predictive maintenance analytics before the system is completely shut down. Once these suspicious changes and slight performance degradation are detected, appropriate action can be performed to mitigate this risk. This will reduce the unnecessary downtime of the machine and increase efficiency by reducing the business loss for manufacturers. Analyzing the historical data of the machines provides insights into their performance and the next scheduled maintenance without affecting their routine task.

B. *Increased Efficiency*: By analyzing and monitoring the data provided by the sensors, IIoT can help manufacturers increase efficiency of their production process by detecting and resolving issues early. This also helps to reduce waste, save energy, increase output, and improve operational efficiency. Additionally, it will help managers by providing accessibility to monitor and control those machines remotely.

C. *Real-Time Visibility with Asset Tracking*: As it will need significant resources to keep track of the large facilities of the industries, applying IIoT helps to automate asset monitoring. RFID tags can be used to track assets or access information during the production state. Furthermore, this information can be transmitted through technologies like Bluetooth and Narrowband IoT (NB-IoT). This gives managers and supervisors information about the situations on the factory floor, helping them to identify problems and their root causes.

D. *Improved Facility Management with Inventory Management*: IIoT can be used in inventory management in which information provided by the sensors will be used to know the status in real time. Information such as inventory levels, rearranging inventory for optimal way, and status of the product such as storage temperature can be managed more efficiently by reducing the operational time. One example of the application of IIoT for inventory management is Amazon, which uses robots that can lift heavy packages, read the labels of the packages using AI and computer vision, and then place the package in the right GoCart for movement within the facilities. This helps to automate these tasks more efficiently, which can be tedious and dangerous for humans.

E. *Save Time and Reduce Costs*: With the use of IIoT in the industry, the overall cost can be reduced by minimizing waste production and increasing the profitability of the business. Additionally, IIoT can identify optimization opportunities by analyzing the data collected during the manufacturing process. This information can help to monitor products during transportation, reducing damage to the items.

F. *Improved Product Quality*: IIoT enables manufacturers to monitor the product along the production line, from raw materials to the finished product. The data gathered by sensors along this line can provide information about the product during the manufacturing process. This can help to detect defects in the product during the manufacturing production line and assist the concerned person in removing those defects. Additionally, the data can be analyzed based on the customer feedback after the product is purchased. This provides information about product quality, customer satisfaction, and issues regarding the product, helping improve its quality.

G. *Safety Control*: IIoT also helps to monitor environments, employee behavior, and compliance for employee safety. This helps to detect hazardous scenarios in the working environment, thus helping to prevent workplace injuries. Equipment such as smart helmets, jackets with sensors, and smart devices provide information about employee status such as movement, body temperature, and perspirations, as well as external environmental conditions. Analyzing those data can help to detect abnormal behaviors like falls, exposure to noxious gases, fires, etc., and notify or alert employees about those conditions. Hence, implementing IIoT has certainly contributed to the workers' safety protocols.

1.6 EMERGING SECTORS IN IIOT

Successful implementation of IIoT requires resources and manpower. Furthermore, other factors in the successful implementation of IIoT in different sectors include boosting computer processing power and communication bandwidth, developing deep learning and ML technology, implementation centers or stations, cloud technology, enhancing the level of public and municipal digital services, and increasing human resource capacity to assist and develop economics. The use of cloud computing in the economy and the enhancement of data availability and quality will determine emerging trends, which are important for the implementation of IIoT in different sectors. The following are some emerging sectors where adoption of IIoT has developed in recent years.

A. *Automotive Industry*: Automotive Industry consists of Connected, semi-autonomous, and autonomous vehicles. IIoT in the automotive industry has enabled vehicles to collect, connect, and exchange real-time data with other smart and intelligent vehicles. These features help to enhance vehicle safety, increase efficiency, and improve user experience. With improved technology and high consumer demand, the adoption and integration of these cutting-edge technologies have spiked in recent years. These technologies also help to achieve autonomous driving, making vehicles smarter and helping in the development of intelligent transportation systems. Analyzing the data provides insights into market dynamics and potential growth, including market share, financial projections, demand and supply ratios, and supply chain efficiencies. It also aids connectivity between vehicles and other devices on the road or wirelessly with other vehicles through the internet. Features such as alerting drivers about traffic conditions, recommending safer routes, and assisting in automatic car parking in tight spots prove beneficial in these vehicles. With advanced connectivity like 5G and AI, auto-driving vehicles are becoming more realistic with the ability to train the model with vast amounts of data gathered from these technologies.

B. *Agriculture Industry*: IIoT in the agriculture sector can greatly benefit farmers by modernizing farming with highly smart machinery and equipment that can gather real-time data. These data can be used to monitor and predict rainfall, soil nutrition levels, and crop yields improving productivity and reducing crop failure. The availability of smart manufacturing devices, such as fertilizer spraying tractors, agricultural drones, automatic agricultural machines, and robots, can help in farming automation without much human effort. IIoT has been implemented in livestock management, smart irrigation, and precision farming. Also, implementing IIoT in agriculture can help in large-scale high-quality production. Recent applications of IIoT in agriculture include drones that can spray pesticides in vast farmland, and collect and analyze data about the production and status of the crops and produce.

C. *Oil and Gas Industry*: Application of IIoT in the oil and gas industry has been for fleet management and route optimization. Information such as oil tanker movement, vehicle conditions, preventive maintenance, driver behavior, and proper monitoring is used for real-time vehicle tracking and route management. Other applications include monitoring the pipeline, leakage, and continuous flow of oil, reducing forced downtime and inevitable expenses. It helps to effectively manage assets and work labor, and to monitor cargo. Using IIoT can provide an effective route for transportation along with smooth supply chain management. It helps to store and manage the oil in the inventory based on the quality of the crude oil and its configuration. Integrating IIoT in the oil and gas industry can help to reduce human labor along with global information with real-time data.

D. *Healthcare*: IIoT in healthcare is one of the most forefront domains which is making high waves in the applications of IIoT. IIoT can be implemented in healthcare sectors to reduce errors and improve overall patient care, healthcare system, and healthcare facilities. Real-time patient authentication [32], patient monitoring using wearables with sensors such as monitoring heart rate, glucose levels, activities, and sleep, can track the patient's health condition. These collected data can be sent to the cloud which can be accessed by all related personnel and healthcare professionals to provide input on ongoing cases [33]. With these technologies, patients' conditions can be monitored automatically and continuously without much human involvement in real time. The system will only alert healthcare professionals if there is something unusual in the collected data. These can also be implemented remotely with much higher efficiency. These may help in detecting health issues early by timely informing healthcare professionals for appropriate medical interventions. However, this adds challenges to avoiding identity theft and preserving the data privacy of patients.

E. *Smart Cities*: With an increasing population that is estimated to reach 9.7 billion by 2050, smart cities and IIoT are much-needed technological developments to improve the resiliency, reliability, management, and security of cities. Several applications that include the use of IoT technology have been developed such as smart lighting, intelligent traffic management, smart waste management technology, environmental monitoring, smart grid and energy management, smart parking management systems, public safety response, health monitoring, and smart building and home automation.

F. *Transportation and Logistics*: The transportation and logistics industry is one of the major sectors where IIoT has been applied and shown great potential for future development [34]. It has revolutionized this industry by using real-time data for optimizing and increasing efficiency in the mobility of goods and services. Its features such as asset tracking, fleet management, traffic management, logistics transport, and supply chain management have benefited the industry to cut down costs, increase efficiency and revenue, decreasing costs for transportation. Apart from this, it has also been applied in other areas such as smart parking, railways, smart toll collections, automatic cars, and predictive maintenance. The real-time data collected from the performance of these assets have helped to increase the energy efficiency, productivity, and reliability of the transportation and logistics industries.

1.7 CHALLENGES IN IIOT

Although applying IIoT has opened new opportunities for businesses to thrive, it has also brought great challenges to the industry. Integrating IoT devices and technology with the legacy manufacturing system has been a challenge in many industries. With established structures, the need to upgrade and advance the application of IoT technologies in manufacturing requires time, effort, and investment, increasing expenses and the need for qualified manpower. With the vast network of interconnected devices, to overcome challenges related to security, the business must ramp up investment in these sectors, which has increased hurdles in the implementation process. This has also increased the concern in companies about the emergence of new risks involved with adopting and applying new technology, which has increased the resistance to change. Combining M2M communications with big data requires uninterrupted connectivity. An outage of connectivity, even during maintenance, can greatly affect data transmission, potentially leading to multiple tragic incidents. Additionally, with the constant monitoring and transferring of data between components, the integrity and confidentiality of these data have become major concerns in the adoption of IIoT in the industry. Hence, the adoption of IIoT brings challenges to the sector such as interoperability, security, privacy, scalability, heterogeneity, reliability, and resource management [35]. The following are some of the challenges faced during the adoption of IIoT.

A. *Security*: One of the main challenges that IIoT faces is in terms of privacy and data security. Most industries have operational processes that require the security and confidentiality of their manufacturing processes, product information, and trade secrets. Information such as product models, personal information of employees, and sensory data should be secured. The adoption of IIoT requires the sharing and communication of these data between several devices. This increases the challenge of securing data and preserving its integrity from other competitive companies. It has also increased the challenge of protecting data from potential external attacks and threats such as phishing. There is an increased need for knowledge of advanced topics in cybersecurity, cyber hygiene, and blockchain implementation to make employees aware of these threats through training and workshops, hereby protecting the companies' networks and assets. Hence, data security, confidentiality, authenticity, and integrity, as well as information about product models and personal information, are significant challenges in the adoption of IIoT in manufacturing companies.

B. *Device Management*: IIoT comprises various technologies and equipment from various vendors and manufacturers that operate in diverse environments. Because of this amalgamation of devices, drafting common established standards for data transfer processes between these different devices, data formats, systems, networks, protocols, and machines has been a challenge in the implementation of IIoT. Also, with the higher potential for scaling IIoT implementation, manufacturing the IIoT devices that support these scalability needs has been a challenge. Manufacturers or vendors may have limited capability to meet the hardware, software, and technical expertise required to scale and implement IIoT effectively. This adds challenges to maintaining the cost and compatibility of upgraded devices with existing devices, affecting the longevity and sustainability of various IIoT technologies.

C. *Patch Management*: With different sensors, actuators, controllers, and robots being used in IIoT technology, these components' software needs to be updated to fix security vulnerabilities, improve performance, and add new features to the system. However, due to the use of different vendors, technologies, or systems, there often are compatibility issues. These different operating systems may use custom software that cannot be easily updated or replaced or use different protocols and architectures, making them incompatible with various patches. This can cause downtime, malfunctions, or damage to the devices or the system, reducing the availability and reliability of the system. Additionally, due to resource constraints, it may be impossible to install, update, or run patches that are dependent on memory, battery power, and connection bandwidth. Hence, patch management has been a challenge as it is important for the performance, security, and functionality of the system.

D. *Cost of Integration*: IIoT systems use different types of sensors, devices, systems, and technologies that have different compatibility and interoperability requirements. Hence, integrating these systems with existing systems and infrastructure will be complex and challenging. Furthermore, integrating systems will need security mechanisms such as encryption, authentication, and authorization to handle sensitive and critical information. This will add extra cost to the system for the protection and communication of the data. Also, maintenance of the integration to ensure that the system is updated for efficient performance requires resources and expertise of manpower that will add additional cost and increase expenses. The lack of common standards, protocols, and platforms compatible with all the IIoT systems, the initial phase of development of technologies such as cloud-based and edge-based solutions, and the lack of integration of automation and AI have increased the cost of the integration and significant challenges for IIoT implementation.

E. *Lack of Standards*: Since IIoT devices use different technologies, protocols, and architecture, they may not be compatible with each other. Additionally, the lack of common standards for all these different technologies has been one of the major challenges in IIoT, hindering interoperability, security, and efficiency of IIoT. Due to the lack of common standards, communication and data exchange between different devices have been difficult which also caused operational disruption and increased vulnerability to cyberattacks. In recent years, there were some emerging standards and protocols that have been developed for IIoT [36] such as MQTT (Message Queuing Telemeter Transport, designed for low-power, low-bandwidth, and unreliable network), OPC UA (Open Platform Communication Unified Architecture, which provides common framework for data exchange and communication), DDS (Data Distribution Service, which provides a global data space without a central broker or server), and OneM2M (which provides common platform for M2M and IoT services). Although there is some development, it still needs more collaboration and coordination among the device manufacturers, service providers, regulators, and end users to develop common standards and protocols to achieve interoperability, security, and efficiency.

F. *Connectivity*: Reliable, secure, and efficient communication between different sensors, devices, and systems for data collection and analysis in real time is one of the key factors for the successful implementation of IIoT. Connectivity can be lost due to technical, environmental, and organizational issues which may affect operational performance, operational cost, and product quality. Due to the performance of the IIoT system being highly dependent on connectivity, flexible, adaptable, and continuous connectivity has been a challenge for the IIoT system. Also, with rapidly growing devices connected to the network each year, the connectivity should be scalable to be able to handle these increased loads and demands to maintain quality and reliability. It should have a large coverage area depending on the locations and environments. Those connections should be highly secure against external threads such as hacking, malware, denial-of-service, or data theft. Data breaches compromise confidentiality and integrity of the data. Furthermore, multiple technologies, protocols, and architecture add to the challenges of making it more compatible with all the technologies and interactions with each other.

G. *Skills*: Implementing IIoT requires skilled manpower for designing, implementing, and managing industrial devices, technology, and systems. However, due to rapid technological change, lack of training and education, and mismatch between supply and demand, it is challenging to find the skill and expertise for efficiently operating the IIoT system. IIoT requires skills such as technical skills related to hardware, software, connectivity, data integration, data analytics, AI, cloud computing, edge computing, and security; operational skills related to functionality, performance, and maintenance of IIoT devices and systems; and business skills related to strategy, value, and impact of IIoT solutions, business analysis, project management, innovation, and management [37]. The rapidly growing market and lack of reskilling and upskilling have added challenges to manage the skilled resources for the successful implementation and management of the IIoT system.

1.8 APPLICATIONS

Applying IIoT has greatly improved industrial automation. It has shown great potential in monitoring and controlling machines and tools using automation without relying solely on humans, hence reducing the impact of labor shortage. Using the IIoT sensors can greatly reduce the labor force needed to check and service manufacturing machines and processes. As these can be implemented remotely, equipment upgrades can be done remotely limiting the direct contact of humans. These can be applied in supply chain, inventory management, and warehouse management systems where efficiency can be increased with this automation process with reduced human oversight. Some of the other sectors where the IIoT process can be applied are as follows:

A. *Automation*: IIoT has a wide range of applications in industrial automation. Automation of machines, equipment, and tools used in the industry increases efficiency and helps to monitor the improvements in the manufacturing process.

   Different tools like Programmable Logic Controllers (PLC) and Programmable Automation Controllers (PAC) are used with sensors to collect data from those IIoT devices in real time. Analyzing the collected data can provide valuable insights that can be used to reduce errors and losses. These machines, which are connected to the network, can be monitored remotely for process flow and downtime and to schedule maintenance.

B. *Production*: IIoT has been applied in production to improve quality and consistency, increase efficiency and productivity, reduce waste and downtime, and optimize flexibility and agility. Devices such as smart sensors, smart actuators, smart cameras, smart robots, and smart software are used to collect data which are used for analyzing and monitoring production conditions, equipment conditions, performance, and product quality. They can be used to optimize production, increase manufacturing speed with automation, and reduce human labor.

C. *Supply Chain*: IIoT has been applied to monitor, track, and optimize the flow of materials, products, and services across the supply chain network. Applying IIoT helps improve visibility and transparency by providing real-time data and insights into the inventory, assets, and shipment information that can be used for better decision-making. Furthermore, it increases efficiency and productivity, reducing costs and risks by enabling predictive analytics that can prevent risks such as inventory shortage, excesses, damages and losses, or transportation and shipping delays. Different state-of-the-art technologies such as RFID tags and readers, GPS trackers and sensors, smart contracts and blockchain, cloud computing, and edge computing provide tracking data which are used for tracking, identifying, and authenticating the inventory, asset, and shipment throughout the supply chain. These technologies can also store, process, and analyze the data from the supply chain devices and systems and use those data to provide scalability, flexibility, and cost-effective computing resources capabilities for supply chain operations.

1.9 CONCLUSION

With the number of smart sensors, smart actuators, and devices increasing in IIoT-connected devices, more data will be collected and analyzed which will create more opportunities as well as challenges in the adoption of IIoT in manufacturing industries in the future. Implementing IIoT can increase production, implement smart manufacturing for higher productivity, optimize supply chain and inventory management, and increase revenue by decreasing cost. Using cloud and edge computing will increase scalability, flexibility, and cost-effectiveness using optimal computing resources. Also, implementing IIoT enables features such as predictive maintenance, data management, real-time location tracking and monitoring, and using blockchain IIoT for supply chain management.

The increasing use and adaptation of IIoT in manufacturing and industrial factories also bring challenges related to security, cost minimization, and integration of technology and systems to be universally compatible, standardized, scalable, and reliable. However, recent technological developments in fields like AI, ML, big data, and edge computing make it capable of automation and provide common platforms. In fields like automation and digital twins, developing intelligent IIoT edge equipment and its implementation holds the rising popularity in IIoT. The future of IIoT includes enhanced device communication, affordable access that will increase productivity, cost savings, and quick detection of issues and opportunities.

REFERENCES

1. K. Ashton and others, "That 'Internet of Things' thing," *RFID J.*, vol. 22, no. 7, pp. 97–114, 2009.
2. R. Alguliyev, Y. Imamverdiyev, and L. Sukhostat, "Cyber-physical systems and their security issues," *Comput. Ind.*, vol. 100, pp. 212–223, 2018.
3. K. P. Jan Stentoft Kent Adsbøll Wickstrøm and A. Haug, "Drivers and barriers for Industry 4.0 readiness and practice: Empirical evidence from small and medium-sized manufacturers," *Prod. Plan. Control*, vol. 32, no. 10, pp. 811–828, 2021, https://doi.org/10.1080/09537287.2020.1768318.
4. V. Vijayaraghavan and J. Rian Leevinson, "Internet of Things applications and use cases in the era of Industry 4.0," In *The Internet of Things in the Industrial Sector: Security and Device Connectivity, Smart Environments, and Industry 4.0*, Z. Mahmood, Ed., Cham: Springer International Publishing, 2019, pp. 279–298, https://doi.org/10.1007/978-3-030-24892-5_12.

5. M. Hermann, T. Pentek, and B. Otto, "Design Principles for Industrie 4.0 Scenarios," In *2016 49th Hawaii International Conference on System Sciences (HICSS)*, 2016, pp. 3928–3937, https://doi.org/10.1109/HICSS.2016.488.

6. J. H. Park, "Advances in future internet and the industrial Internet of Things," *Symmetry (Basel)*, vol. 11, no. 2, pp. 24, 2019, https://doi.org/10.3390/sym11020244.

7. K. Okundaye, S. K. Fan, and R. J. Dwyer, "Impact of information and communication technology in Nigerian small-to medium-sized enterprises," *J. Econ. Financ. Adm. Sci.*, vol. 24, no. 47, pp. 29–46, 2019, https://doi.org/10.1108/JEFAS-08-2018-0086.

8. J. Lee, B. Bagheri, and H.-A. Kao, "A cyber-physical systems architecture for Industry 4.0-based manufacturing systems," *Manuf. Lett.*, vol. 3, pp. 18–23, 2015.

9. S. R. Shakya and S. Jha, "Chapter 9- Internet of Things application in life sciences," In *Internet of Things in Biomedical Engineering*, V. E. Balas, L. H. Son, S. Jha, M. Khari, and R. Kumar, Eds., London, UK: Academic Press, 2019, pp. 213–233, https://doi.org/10.1016/B978-0-12-817356-5.00012-7.

10. M. H. ur Rehman, I. Yaqoob, K. Salah, M. Imran, P. P. Jayaraman, and C. Perera, "The role of big data analytics in industrial Internet of Things," *Futur. Gener. Comput. Syst.*, vol. 99, pp. 247–259, 2019, https://doi.org/10.1016/j.future.2019.04.020.

11. J. Lin, W. Yu, N. Zhang, X. Yang, H. Zhang, and W. Zhao, "A survey on internet of things: Architecture, enabling technologies, security and privacy, and applications," *IEEE Internet Things J.*, vol. 4, no. 5, pp. 1125–1142, 2017.

12. H. Xu, W. Yu, D. Griffith, and N. Golmie, "A survey on industrial Internet of Things: A cyber-physical systems perspective," *IEEE Access Pract. Innov. Open Solut.*, vol. 6, pp. 78238–78259, 2018, https://doi.org/10.1109/access.2018.2884906.

13. M. A. Khan, "A survey of security issues for cloud computing," *J. Netw. Comput. Appl.*, vol. 71, pp. 11–29, 2016, https://doi.org/10.1016/j.jnca.2016.05.010.

14. H. Boyes, B. Hallaq, J. Cunningham, and T. Watson, "The Industrial Internet of Things (IIoT): An analysis framework," *Comput. Ind.*, vol. 101, pp. 1–12, 2018, https://doi.org/10.1016/j.compind.2018.04.015.

15. A. A. Alli and M. M. Alam, "The fog cloud of things: A survey on concepts, architecture, standards, tools, and applications," *Internet of Things*, vol. 9, p. 100177, 2020, https://doi.org/10.1016/j.iot.2020.100177.

16. H. Huang, L. Yang, Y. Wang, X. Xu, and Y. Lu, "Digital Twin-driven online anomaly detection for an automation system based on edge intelligence," *J. Manuf. Syst.*, vol. 59, pp. 138–150, 2021, https://doi.org/10.1016/j.jmsy.2021.02.010.

17. P. Zhen, Y. Han, A. Dong, and J. Yu, "CareEdge: A lightweight edge intelligence framework for ECG-based heartbeat detection," *Procedia Comput. Sci.*, vol. 187, pp. 329–334, 2021, https://doi.org/10.1016/j.procs.2021.04.070.

18. L. Mas, J. Vilaplana, J. Mateo, and F. Solsona, "A queuing theory model for fog computing," *J. Supercomput.*, vol. 78, no. 8, pp. 11138–11155, 2022, https://doi.org/10.1007/s11227-022-04328-3.

19. M. H. ur Rehman, V. Chang, A. Batool, and T. Y. Wah, "Big data reduction framework for value creation in sustainable enterprises," *Int. J. Inf. Manage.*, vol. 36, no. 6, Part A, pp. 917–928, 2016, https://doi.org/10.1016/j.ijinfomgt.2016.05.013.

20. E. G. Nabati and K.-D. Thoben, "Data driven decision making in planning the maintenance activities of off-shore wind energy," *Procedia CIRP*, vol. 59, pp. 160–165, 2017, https://doi.org/10.1016/j.procir.2016.09.026.

21. F. Raheem and N. Iqbal, "Artificial intelligence and machine learning for the Industrial Internet of Things (IIoT)," In *Industrial Internet of Things*, Boca Raton, FL: CRC Press, 2022, pp. 1–20.

22. W. Jiang, H. Li, G. Xu, M. Wen, G. Dong, and X. Lin, "PTAS: Privacy-preserving Thin-client authentication scheme in blockchain-based PKI," *Futur. Gener. Comput. Syst.*, vol. 96, pp. 185–195, 2019, https://doi.org/10.1016/j.future.2019.01.026.

23. M. Shen, X. Tang, L. Zhu, X. Du, and M. Guizani, "Privacy-preserving support vector machine training over blockchain-based encrypted IoT data in smart cities," *IEEE Internet Things J.*, vol. 6, no. 5, pp. 7702–7712, 2019, https://doi.org/10.1109/JIOT.2019.2901840.

24. M. Ma, G. Shi, and F. Li, "Privacy-oriented blockchain-based distributed key management architecture for hierarchical access control in the IoT scenario," *IEEE Access*, vol. 7, pp. 34045–34059, 2019, https://doi.org/10.1109/ACCESS.2019.2904042.

25. L. Zhou, L. Wang, Y. Sun, and P. Lv, "BeeKeeper: A blockchain-based IoT system with secure storage and homomorphic computation," *IEEE Access*, vol. 6, pp. 43472–43488, 2018, https://doi.org/10.1109/ACCESS.2018.2847632.

26. Y. Lin et al., "A novel architecture combining oracle with decentralized learning for IIoT," *IEEE Internet Things J.*, vol. 10, no. 5, pp. 3774–3785, 2023, https://doi.org/10.1109/JIOT.2022.3150789.

27. F. Ghovanlooy Ghajar, A. Sikora, and D. Welte, "Schloss: Blockchain-based system architecture for secure industrial IoT," *Electronics*, vol. 11, no. 10, 2022, https://doi.org/10.3390/electronics11101629.

28. "5 G-ACIA. 5G Alliance for Connected Industries and Automation." [Online]. Available: https://5g-acia.org/.

29. S. K. Rao and R. Prasad, "Impact of 5G technologies on Industry 4.0," *Wirel. Pers. Commun.*, vol. 100, no. 1, pp. 145–159, 2018.

30. Z. Yang, J. He, Y. Tian, and J. Zhou, "Faster authenticated key agreement with perfect forward secrecy for industrial Internet-of-Things," *IEEE Trans. Ind. Inform.*, vol. 16, no. 10, pp. 6584–6596, 2019, https://doi.org/10.1109/TII.2019.2963328.

31. S. R. Chhetri, N. Rashid, S. Faezi, and M. A. Al Faruque, "Security trends and advances in manufacturing systems in the era of Industry 4.0," In *2017 IEEE/ACM International Conference on Computer-Aided Design (ICCAD)*, 2017, pp. 1039–1046. https://doi.org/10.1109/ICCAD.2017.8203896.

32. F. Al-Turjman and S. Alturjman, "Context-sensitive access in industrial Internet of Things (IIoT) healthcare applications," *IEEE Trans. Ind. Inform.*, vol. 14, no. 6, pp. 2736–2744, 2018, https://doi.org/10.1109/TII.2018.2808190.

33. M. S. Hossain and G. Muhammad, "Cloud-assisted Industrial Internet of Things (IIoT) - Enabled framework for health monitoring," *Comput. Netw.*, vol. 101, pp. 192–202, 2016, https://doi.org/10.1016/j.comnet.2016.01.009.

34. Y. Zhang, Z. Guo, J. Lv, and Y. Liu, "A framework for smart production-logistics systems based on CPS and industrial IoT," *IEEE Trans. Ind. Inform.*, vol. 14, no. 9, pp. 4019–4032, 2018, https://doi.org/10.1109/TII.2018.2845683.

35. S. R. Shakya and S. Jha, "Challenges in Industrial Internet of Things (IIoT)," In *Industrial Internet of Things*, Boca Raton, FL: CRC Press, 2022, pp. 19–39.
36. Y. Lu, P. Witherell, and A. Jones, "Standard connections for IIoT empowered smart manufacturing," *Manuf. Lett.*, vol. 26, pp. 17–20, 2020, https://doi.org/10.1016/j.mfglet.2020.08.006.
37. I. Lee and K. Lee, "The Internet of Things (IoT): Applications, investments, and challenges for enterprises," *Bus. Horiz.*, vol. 58, no. 4, pp. 431–440, 2015, https://doi.org/10.1016/j.bushor.2015.03.008.

Chapter 2

Security challenges in the Industrial Internet of Things (IIoT)

*Saydul Akbar Murad, Nick Rahimi,
Indranil Roy, and Bidyut Gupta*

2.1 INTRODUCTION

The Industrial Internet of Things (IIoT) has brought about a new era of industrialization that has made manufacturing more efficient, productive, and cost-effective [1]. However, with the growing popularity of IIoT, security concerns have also grown, making the need for secure IIoT systems more critical than ever before. Cyber-attacks, data breaches, device vulnerabilities, and insider threats are just a few security challenges that IIoT systems face [1]. According to a report by Symantec, the number of attacks on IIoT systems increased by 220% in 2017 alone [1]. This highlights the urgent need to develop effective security mechanisms to protect IIoT systems against various threats. This chapter examines the security challenges in IIoT and the computer science-related techniques that can mitigate those challenges.

This chapter discusses the specific security challenges in IIoT including cyber-attacks, data breaches, device vulnerabilities, and insider threats. This will be followed by an overview of security, privacy, confidentiality, blockchain, and machine-learning techniques that can be used to mitigate security challenges. Finally, the chapter concludes by summarizing the key points and emphasizing the importance of IIoT security.

2.2 SECURITY CHALLENGES IN IIOT

IIoT has created a connected network of devices, machines, and sensors that communicate and share data (see Figure 2.1 for illustration). While this has brought about several benefits, it has also created new security challenges [2]. This section discusses the major security challenges IIoT systems face, including cyber-attacks, data breaches, device vulnerabilities, and insider threats.

DOI: 10.1201/9781003530572-2

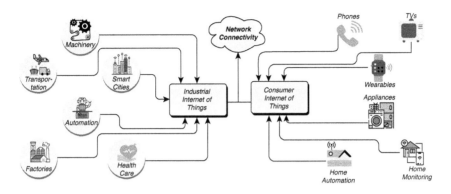

Figure 2.1 Illustration of IIoT.

2.2.1 Cyber-attacks

Cyber-attacks are one of the most significant security challenges facing IIoT systems [3]. Attackers can target IIoT systems with various attack methods, including malware, ransomware, and denial-of-service (DoS) attacks. One of the most well-known cyber-attacks on an IIoT system is the Stuxnet worm attack, which targeted Iran's nuclear program in 2010. Stuxnet was a sophisticated worm that targeted industrial control systems (ICSs) and caused physical damage to the centrifuges in Iran's nuclear program [4]. In addition to Stuxnet, other cyber-attacks on IIoT systems have targeted critical infrastructure, such as power grids and water treatment facilities. The risk of DoS attacks cannot be overlooked when it comes to IIoT security, particularly due to the low-power and low-capacity nature of IIoT edge devices. Networks lacking adequate authentication and intrusion detection mechanisms are particularly vulnerable to such attacks, which can cause significant disruption and damage to IIoT systems [3]. Figure 2.2 illustrates a DoS attack using an RFID reader.

2.2.2 Data breaches

According to Byres [2], data breaches are another significant security challenge facing IIoT systems. As IIoT devices collect and transmit sensitive data, such as industrial secrets and proprietary information, the risk of data breaches increases. Hackers can exploit vulnerabilities in IIoT devices to gain unauthorized access to sensitive data, resulting in financial losses, reputational damage, and legal liabilities. For instance, in 2018, a data breach at a steel mill in Germany resulted in significant financial losses due to stolen trade secrets [5].

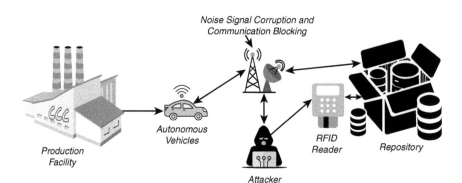

Figure 2.2 Illustration of DoS attack in IIoT.

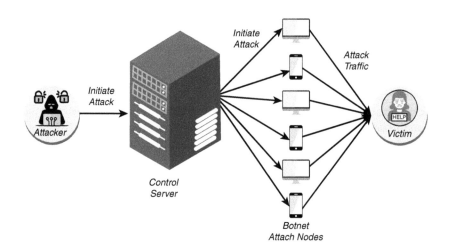

Figure 2.3 Illustration of DDoS Botnet.

2.2.3 Device vulnerabilities

Device vulnerabilities are another security challenge facing IIoT systems [1,6]. As IIoT devices become more interconnected, the attack surface increases, making it easier for attackers to exploit vulnerabilities in the devices. In addition, many IIoT devices are designed with limited processing power and memory, which can make it challenging to implement effective security measures [3].

One example of device vulnerabilities in IIoT systems is the Mirai Botnet attack, which targeted Internet of Things (IoT) devices, including cameras, routers, and digital video recorders (DVRs) [7]. The attackers exploited a vulnerability in the devices' default login credentials, allowing them to take control of the devices and use them to launch distributed denial-of-service (DDoS) attacks. Figure 2.3 illustrates the Mirai Botnet attack. By using a compromised system, an attacker can increase the scope of the attack.

2.2.4 Insider threats

In the view of Bajramovic et al. [8], insider threats are a lesser-known but equally significant security challenge facing IIoT systems. Insiders like employees and contractors have access to sensitive data and systems, potentially threatening IIoT security. Karakaya and Arat [3] assert that insiders can intentionally or unintentionally cause security breaches by misusing their access privileges, mishandling sensitive data, or failing to follow security policies. For instance, in 2019, an employee of Tesla was accused of stealing sensitive data, including Autopilot source code, and sharing it with outsiders.

2.3 TECHNIQUES TO MITIGATE SECURITY CHALLENGES

Different techniques can be used to mitigate security challenges in IIoT. They can be broadly classified as security techniques, privacy techniques, and confidentiality techniques.

2.3.1 Security techniques

IIoT presents numerous security challenges due to the interconnected nature of its devices, data, and systems [8,9]. Tan and Samsudin [1] argue that to ensure the integrity, confidentiality, and availability of IIoT systems, it is necessary to employ various security techniques, as discussed here.

2.3.1.1 Encryption

According to Yu and Guo [10], encryption is a crucial security technique in IIoT systems, as it ensures that data transmitted over the network is protected from unauthorized access. Encryption involves transforming plain text data into cipher text using an encryption algorithm (see Figure 2.4) and a secret key [2]. Someone with the secret key can only decrypt the cipher text. Yu and Guo [10] assert that end-to-end encryption is recommended to protect data in IIoT systems. It ensures that the data is encrypted at the source and can only be decrypted by the intended recipient, eliminating the risk of interception or eavesdropping.

2.3.1.2 Access control

Access control is an essential security technique in IIoT systems, as it ensures that only authorized users can access the system, devices, or data [3,11]. Access control is implemented through authentication, authorization, and accounting. Authentication involves verifying the identity of a user or device before granting access. This can be done using various methods, such as password-based authentication, biometric authentication, or multi-factor authentication [8]. Authorization involves granting specific

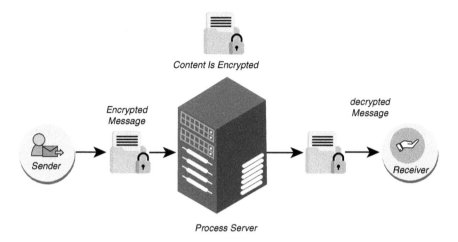

Figure 2.4 End-to-end encryption.

permissions to a user or device based on their role and level of access. For example, an administrator may have full access to the IIoT system, while an operator may only access specific devices. Accounting involves keeping track of the actions performed by users or devices within the IIoT system. This ensures that any unauthorized actions can be traced back to the user or device responsible [1].

2.3.1.3 Authentication

According to Tan and Samsudin [1], authentication is a crucial security technique in IIoT systems, as it ensures that only authorized users or devices can access the system. Password-based authentication is the most common method of authentication in IIoT systems. However, passwords can be vulnerable to various attacks, such as brute force attacks or password guessing attacks [10]. Malti [12] supports biometric authentication that involves using physiological or behavioral characteristics to verify a user's or device's identity, such as fingerprint or facial recognition. Also, multi-factor authentication, characterized by using multiple authentication methods to enhance the system's security, such as combining a password with a fingerprint scan, is highly recommended.

2.3.1.4 Intrusion detection and prevention

Intrusion detection and prevention systems (IDPS) monitor IIoT systems for suspicious activities and detect and prevent potential threats. IDPS can use various techniques such as signature-based detection, anomaly-based detection, or behavior-based detection to identify potential threats and prevent

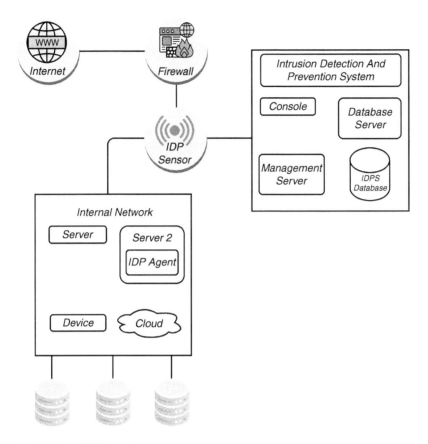

Figure 2.5 How IDPS works.

them from compromising IIoT systems [10] (see Figure 2.5). Signature-based detection involves matching incoming traffic against a database of known attack signatures. Anomaly-based detection includes identifying behavior patterns outside the norm and flagging them as potentially suspicious [2,13]. Behavior-based detection involves monitoring the behavior of users or devices within the system and identifying potential threats based on deviations from the norm [12].

2.3.2 Privacy techniques

According to Malti [10], privacy is essential to IIoT systems, as they collect and process vast amounts of data, including personal and sensitive information. As such, privacy techniques are used to protect this information from unauthorized access, use, or disclosure. One of the main privacy techniques is anonymization, which involves removing or obscuring any identifiable information from the data, making it impossible to link the data to a specific individual or device [3]. In the view of Malti [10], anonymization is commonly used in IIoT systems to protect personal information and allow the data to be analyzed for insights or trends. Tan and Samsudin [1] argue that anonymization can be achieved through various methods, such as generalization, randomization, or hashing. However, Byres [2] opines that it is important to note that complete anonymization can be difficult to achieve, as it requires removing all identifying information from the data, which can affect the accuracy and usefulness of the data.

Accordingly, Bajramovic [8] argues that pseudonymization is another privacy technique that replaces identifiable information with a pseudonym. Pseudonym is a unique identifier that cannot be traced back to the original data subject without additional information. Pseudonymization is a useful technique in IIoT systems, as it allows the data to be analyzed and protects the privacy of the data subjects. Pseudonymization can be achieved through various methods, such as encryption or tokenization [8].

Differential privacy is another privacy technique that involves adding random noise to the data to prevent individual data points from being identified [3]. This technique is used to protect the privacy of individuals and ensure data is correctly analyzed. Differential privacy is commonly used in IIoT systems that collect and analyze sensitive information, such as healthcare or financial data [10].

2.3.3 Confidentiality techniques

Malti [10] argues that confidentiality is critical to IIoT systems, as they often handle sensitive information that should not be accessible to unauthorized individuals or entities. According to Karakaya and Arat [3], confidentiality techniques are used to protect the confidentiality of data and prevent unauthorized access or disclosure. One of the confidentiality techniques is secret sharing, which involves splitting the data into multiple shares and distributing them among different parties. The data can only be reconstructed by combining the shares held by all parties [3]. Secret sharing is commonly used in IIoT systems to protect sensitive data, such as passwords or encryption keys, from being accessed or stolen by a single party. This technique can be implemented using various methods, such as Shamir's secret or threshold secret sharing [8,14]. Another technique is homomorphic encryption, which allows computations to be performed on encrypted data without decrypting it first. Homomorphic encryption can be implemented using various methods, such as fully or partially homomorphic. Malti [10] argues that these confidentiality techniques can be combined with security and privacy techniques to ensure the confidentiality, privacy, and security of IIoT systems.

2.4 BLOCKCHAIN IN IIOT SECURITY

Blockchain technology has gained widespread attention recently due to its potential applications in various industries. One industry that could greatly benefit from blockchain technology is the IIoT sector [15,16]. According to Sengupta et al. [16], through IIoT, various devices and machines in an industrial setting are interconnected, where data is constantly being generated and transmitted across various networks. This has led to massive amounts of data that must be processed and analyzed in real time. However, this also poses a significant security risk, as any unauthorized access or tampering with the data can have severe consequences [10,16]. Blockchain technology, essentially a distributed ledger system that is immutable and decentralized, can potentially address many security concerns associated with IIoT [16]. According to Zhao and Yao [17], one of the key advantages of blockchain technology is its ability to create a tamper-proof record of data transactions. This means that once data is entered into the blockchain, it cannot be altered or deleted, ensuring its integrity.

Furthermore, Zhao and Yao [17] add that since the blockchain is a distributed ledger system, it allows a decentralized network of nodes to maintain and validate the data. This eliminates the need for a central authority to control the data, reducing the risk of a single point of failure. In the context of IIoT, this means that data can be securely transmitted across a network of devices and machines without the need for a centralized intermediary [18]. Additionally, Choo, Yan, and Meng [19] argue that blockchain technology is its ability to provide transparency and audibility of data transactions. Since all data transactions are recorded on the blockchain, it is possible to trace the entire history of a particular data point, enhancing accountability and enabling better data auditing.

A study by Liu et al. [20] explored the use of blockchain technology to secure IIoT systems by creating a decentralized security framework. The study highlighted the advantages of using blockchain technology, such as its ability to provide data integrity, immutability, and transparency, which are critical in securing IIoT infrastructure. Similarly, a study by Wu et al. [21] discussed the use of blockchain technology to improve the security of IIoT networks. The study proposes a blockchain-based framework that utilizes smart contracts to ensure data confidentiality, integrity, and availability. The authors argue that such a framework can significantly enhance the security of IIoT systems by providing a secure and decentralized platform for data transactions.

In summary, Wu et al. [21] assert that blockchain technology has the potential to significantly enhance the security of IIoT systems by providing a tamper-proof and decentralized platform for data transactions. The use of blockchain technology can improve data integrity, transparency, and audibility, which can enhance accountability and reduce the risk of unauthorized access or tampering [10,18,19]. As IIoT systems become increasingly prevalent in various industries, blockchain technology will likely become more widespread, and research in this area will likely continue to expand.

2.5 MACHINE LEARNING IN IIOT SECURITY

Machine learning (ML) is among the most sought-after technologies radically changing how data is extracted and interpreted. In IIoT security, ML has significant potential to improve the detection and mitigation of security threats [8]. In IIoT, massive amounts of data need to be processed and analyzed in real time. ML algorithms can analyze this data to detect and prevent security threats [10].

Wu et al. [21] argue that ML in IIoT security can identify patterns and anomalies in data. This means that ML algorithms can be trained to recognize unusual behavior and detect potential security threats promptly. For example, ML algorithms can detect abnormal network activity, such as unauthorized access or data exfiltration, which can be a sign of security breach. Aris et al. [22] also argue that ML in IIoT security can adapt and learn from new data. ML algorithms can be trained to recognize new patterns and adjust their models accordingly as new data is generated. This means that ML algorithms can evolve and become more effective at detecting security threats.

Shahid et al. [23] explored the use of ML algorithms to detect cyber-attacks in IIoT networks and found that ML algorithms are effective in detecting various types of cyber-attacks, such as DDoS attacks and malware infections. Similarly, a study by Sengupta [16] found that ML algorithms enhance the security of IIoT networks. The study proposed an ML-based framework that uses various data sources, such as network traffic and system logs, to detect and prevent security threats.

In a nutshell, ML has significant potential to improve the detection and mitigation of security threats in IIoT networks. ML algorithms can be trained to detect patterns and anomalies in data, which can be used to detect potential security threats in real time [19,18,23,24]. As IIoT systems become increasingly prevalent in various industries, ML in IIoT security will likely become more widespread, and research in this area will continue to expand.

2.6 CONCLUSION

In conclusion, this chapter has provided an overview of the security challenges in IIoT and discussed various techniques that can be used to mitigate these challenges. The importance of IIoT security cannot be overstated, as IIoT systems play a critical role in various industries, and a security breach can result in severe consequences. Therefore, it is crucial to implement appropriate security measures, including encryption, access control, authentication, intrusion detection and prevention, privacy techniques, blockchain, and ML. However, IIoT security is an ongoing process that requires continuous monitoring and adaptation to stay ahead of evolving cyber threats. As IIoT systems continue to evolve and become more inter-connected, it is essential to prioritize security and implement robust secu-rity measures to protect against cyber threats.

REFERENCES

1. T. S. Fun and A. Samsudin, "Recent technologies, security countermeasure and ongoing challenges of industrial internet of things (IIoT): A survey," *Sensors*, vol. 21, no. 19, pp. 6647, 2021. https://doi.org/10.3390/s21196647.
2. E. J. Byres, "Companies need to change focus, mindset on IIoT security: There are ways for companies to get an Industrial Internet of Things (IIoT) project initiated while overcoming," *E J Byres Control Eng. 2017 go.gale. com* [Online]. Available: https://go.gale.com/ps/i.do?id=GALE%7CA48908 1466&sid=googleScholar&v=2.1&it=r&linkaccess=abs&issn=00108049&p= AONE&sw=w (accessed: Jan. 18, 2024).
3. H. K. Saini, M. Poriye, and N. Goyal, "A survey on security threats and network vulnerabilities in Internet of Things," *International Journal of Information Security Science*, vol. 10, no. 4, pp. 297–314, 2024. https://doi. org/10.1007/978-981-99-4518-4_18.
4. Paulius Ilevičius, "Stuxnet explained – the worm that went nuclear | NordVPN," *NordVPN*, 2022. https://nordvpn.com/blog/stuxnet-virus/ (accessed Jan. 18, 2024).
5. BBC News, "Hack attack causes 'massive damage' at steel works," *BBC News*, 2014. https://www.bbc.com/news/technology-30575104 (accessed Jan. 18, 2024).
6. N. Rahimi and S. S. V. Gudapati, "Emergence of blockchain technology in the healthcare and insurance industries," *Blockchain Technology Solutions for the Security of IoT-Based Healthcare Systems*, vol. 2023, pp. 167–182, 2023. https://doi.org/10.1016/B978-0-323-99199-5.00013-6.
7. Cloudflare, "What is the Mirai Botnet? | Cloudflare," 2022. https://www. cloudflare.com/learning/ddos/glossary/mirai-botnet/ (accessed Jan. 18, 2024).
8. E. Bajramovic, D. Gupta, Y. Guo, K. Waedt, and A. Bajramovic, "Security challenges and best practices for IIoT," *Lecture Notes in Informatics (LNI), Proceedings – Series of the Gesellschaft fur Informatik (GI)*, vol. 294, no. 9, pp. 125–138, 2019, https://doi.org/10.18420/inf2019_18.
9. S. A. Murad and N. Rahimi, "Secure and scalable permissioned block-chain using LDE-P2P networks," *2023 10th International Conference on Internet of Things: Systems, Management and Security (IOTSMS 2023) San Antonio, Texas , USA*, pp. 111–116, 2023, https://doi.org/10.1109/ IOTSMS59855.2023.10325762.
10. M. Bansal, A. Goyal, and A. Choudhary, "Industrial Internet of Things (IIoT): A vivid perspective," *Lecture Notes in Networks and Systems*, vol. 204, pp. 939–949, 2021. https://doi.org/10.1007/978-981-16-1395-1_68.

11. S. A. Murad, N. Rahimi, I. Roy, and B. Gupta, "Design of a new secured hierarchical peer-to-peer fog architecture based on linear diophantine equation," *Epic Series Computing*, vol. 91, pp. 86–97, 2023. https://doi.org/10.29007/qn8f.

12. M. Ramya, "What is intrusion detection and prevention system? Definition, examples, techniques, and best practices – Spiceworks," *Spiceworks*, 2022. https://www.spiceworks.com/it-security/vulnerability-management/articles/what-is-idps/ (accessed Jan. 18, 2024).

13. N. Rahimi, "A study of the landscape of security issues, vulnerabilities, and defense mechanisms in web based applications," in *Proceedings – 2021 International Conference on Computational Science and Computational Intelligence, CSCI 2021*, pp. 806–811, 2021. https://doi.org/10.1109/CSCI54926.2021.00194.

14. N. Rahimi, I. Roy, B. Gupta, P. Bhandari, and N. C. Debnath, "Blockchain technology and its emerging applications," *Blockchain Technology for Data Privacy Management*, vol. 2021, pp. 133–157, 2021. https://doi.org/10.1201/9781003133391-7.

15. N. Rahimi and B. Gupta, "Security issues, vulnerabilities, and defense mechanisms in wireless sensor networks: State of the art and recommendation," in *Integration of WSNs into Internet of Things*, CRC Press, Boca Raton, FL, pp. 1–15, 2021. https://doi.org/10.1201/9781003107521-1.

16. J. Sengupta, S. Ruj, and S. Das Bit, "A comprehensive survey on attacks, security issues and blockchain solutions for IoT and IIoT," *Journal of Network and Computer Applications*, vol. 149, pp. 102481, 2020. https://doi.org/10.1016/j.jnca.2019.102481.

17. S. Zhao, S. Li, and Y. Yao, "Blockchain enabled industrial Internet of Things technology," *IEEE Transactions on Computational Social Systems*, vol. 6, no. 6, pp. 1442–1453, 2019. https://doi.org/10.1109/TCSS.2019.2924054.

18. D. Job, and V. Paul, "Challenges, security mechanisms, and research areas in IoT and IIoT," *Springer Internet Things Its Applications EAI 2022*. Springer, New York, 2022 [Online]. Available: https://link.springer.com/chapter/10.1007/978-3-030-77528-5_28 (accessed: Jan. 18, 2024.)

19. K. K. R. Choo, Z. Yan, and W. Meng, "Editorial: Blockchain in industrial IoT applications: Security and privacy advances, challenges, and opportunities," *IEEE Transactions on Industrial Informatics*, vol. 16, no. 6. pp. 4119–4121, 2020. https://doi.org/10.1109/TII.2020.2966068.

20. X. Liu, Y. Yang, K. K. R. Choo, and H. Wang, "Security and privacy challenges for internet-of-things and fog computing," *Wireless Communications and Mobile Computing*, vol. 2018, pp. 34–42, 2018. https://doi.org/10.1155/2018/9373961.

21. Y. Wu, H. N. Dai, and H. Wang, "Convergence of blockchain and edge computing for secure and scalable IIoT critical infrastructures in Industry 4.0," *IEEE Internet Things Journal*, vol. 8, no. 4, pp. 2300–2317, 2021. https://doi.org/10.1109/JIOT.2020.3025916.

22. X. Yu and H. Guo, "A survey on IIoT security," in *Proceedings of 2021 IEEE VTS 17th Asia Pacific Wireless Communications Symposium APWCS 2019*, 2019, https://doi.org/10.1109/VTS-APWCS.2019.8851679.

23. A. S. Lalos, A. P. Kalogeras, C. Koulamas, C. Tselios, C. Alexakos, and D. Serpanos, "Secure and safe IIoT systems via machine and deep learning approaches," in *Security and Quality in Cyber-Physical Systems Engineering*, Springer International Publishing, New York, pp. 443–470, 2019. https://doi.org/10.1007/978-3-030-25312-7_16.
24. N. Rahimi, B. Gupta, and S. Rahimi, "Secured data lookup in LDE based low diameter structured P2P network," in *Proceedings of the 33rd International Conference on Computers and Their Applications, CATA 2018*, 2018, vol. 2018-March. [Online]. Available: https://www.cse.msstate.edu/wp-content/uploads/2020/04/I6_rahimi.pdf (accessed: Jan. 18, 2024).

Chapter 3

Optimizing quality control in IIoT-based manufacturing

Leveraging big data analytics
and IoT devices for enhanced
decision-making strategies

Gauri Mathur, Raj Karan Singh,
Manik Rakhra, and Deepak Prashar

INTRODUCTION

Nowadays, every sector uses the Internet of Things (IoT) for transmitting their valuable data from one machine to another. IoT technology is used by several industries, including retail and resource management, manufacturing and logistics, traffic control, IT companies, and hospital management, to collect data through a network of connected devices. Examples of these devices include delivery trucks, drones, medical equipment, CCTV cameras, and construction equipment. IoT devices [1] and sensors have the ability to gather and extract millions of useful data points, producing massive, fast-moving data streams that are difficult to store, handle, analyze, and secure. Since this data is highly perishable, it must be extracted with the right instruments and technology to avoid missing opportunities for enterprises to act on sensitive inputs vital to their operations. Data streams can capture real-time data transmission and then send it to the cloud for offline processing. We use various technologies such as IoT platforms like Kaa and AWS, 5G, and analytics tools like Kafka, Kinesis, Spark, Storm, Cassandra, and BigTable for real-time data analytics. The hardware, software, and other equipment used for real-time data transmission is very affordable, leading to rapid growth in the number of connected devices used to capture data streams such as audio, video, and images of varied nature.

IoT and big data analytics are emerging as important tools in today's competitive world. Big IoT data analytics give businesses the ability to extract useful information from IoT sensors, devices, and third-party datasets for analysis using current business tools. This important information can be used to improve information, services, and products, as well as to improve decision-making processes. However, all organizations need the necessary infrastructure in place to handle data transactions in real time.

DOI: 10.1201/9781003530572-3

3.1 OVERVIEW OF IOT AND BIG DATA

3.1.1 IoT networks

IoT [2] provides one of the best mediums for sensor devices and other wireless gadgets to communicate effortlessly in a smart and disturbance-free environment and also helps in sharing information from one device to another device without infrastructure requirements. IoT networking is one of the most promising fields and emerging trends where every device, whether in the field of smart cities, banking, military, transportation, or medical fields can exchange data with each other without any pre-set infrastructure. Every device that is being used in day-to-day life like mobile phones, TVs, electronic appliances like microwaves, geysers, laptops, washing machines, emergency alarms, and vending machines can be connected via an IoT network and can be remotely controlled. These devices can talk to each other and can share and collect various sorts of data like computational data, geographical data, environmental data, and logical data with each other.

In the IoT paradigm, a significant number of these devices [3] are integrated with real-world sensor equipment. After sensing the data, the data fetching nodes use embedded electrical systems to convey the data to other devices. These gadgets, along with other embedded systems, are interconnected by a variety of communication channels, including WiFi, GSM, Bluetooth, and ZigBee. These devices have two functions: they can send and receive data via remotely operated devices. They can also integrate technology with physical devices and gadgets to raise people's standards of living. More than 60 billion devices, including gaming consoles, laptops, smartphones, sensors, and electronic equipment, are encouraged to connect to the internet via a variety of heterogeneous access networks made possible by technologies like wireless sensor networks and radio frequency identification (RFID).

Three criteria are used to classify the IoT: knowledge, sensors, and internet-orientedness. The adoption of this technology by several smart devices has improved people's quality of life and allowed daily tasks to be completed flawlessly and problem-free.

3.1.2 Big data

Large volumes of structured, unstructured, and semi-structured data are produced by sensors, cellphones, laptops, and applications such as banking, temperature monitoring, and medical. These applications continuously produce enormous volumes of data that can be challenging to manage. So, big data refers to the enormous volume of data produced by devices. The vast volumes of data produced by these systems cannot be efficiently stored, processed, or analyzed using traditional database management techniques.

Thus, the creation of big data is highly sought after in the business and IT sectors these days. An instance of research pertaining to big data is the next frontier in terms of productivity, competition, and innovation. According to the study "The Digital Universe," big data technologies are considered to be of the current generation. Their extensive architecture allows for the extraction of pertinent information from the massive amounts of data produced by these devices, which can be advantageous for businesses as it supports strategic decision-making and helps with long-term planning. Big data is divided into three categories: data sources, data analytics, and result display. It is predicated on Beyer's 3Vs paradigm, which stands for volume, velocity, and variety. These are the main components that determine how big data produced by different devices is measured. The device's primary characteristic, volume [3], is where a huge amount of data is continuously generated. With the aid of various big data analytics methods, we can extract only useful data. Variety refers to the range of data types that these devices produce and the speed at which they do so. These data types must be cleaned, refined, separated, and ultimately transformed into meaningful information.

3.2 USAGE OF BIG DATA IN IOT

Big data and IoT are developing quickly and are being used in every aspect of business and technology to improve outcomes for both individuals and industries. Recent advancements have been greatly influenced by the amount of data produced by IoT devices. Three factors are used to categorize big data [4]: volume, diversity, and velocity. The ability to evaluate and make use of the massive volumes of data created by IoT devices, such as those in smart cities, smart gadgets, smart transportation systems, and remote patient healthcare monitoring devices, presents a plethora of options.

Because so many different sensor devices are used in the IoT infrastructure for data collection and processing, big data analytics has become increasingly difficult due to the massive rise in the popularity of IoT. Big data analytics for IoT is defined as the critical phase and procedures whereby different types of IoT data are examined and processed to find new correlations, hidden information, and unexpected trends. Many individuals and companies can benefit greatly from analyzing large volumes of data and organizing large quantities of useful information to advance business. IoT big data analytics is likely to constantly support companies and business partnerships in their efforts to gain a deeper comprehension of the vast amounts of data created by these devices to make more precise and effective decisions. Large amounts of data are produced by analytics as a result of organizing and refining unstructured data, which can benefit businesses and facilitate improved decision-making. Big data analytics, on the other hand, emphasizes the extraction of knowledgeable information through a variety of data mining approaches, which aids in enhancing predictions, uncovering information that is concealed, identifying current trends, and making wise decisions.

Several problems with processing, storing, visualizing, and data analytics can be helped by integrating big data and IoT technologies. In a smart city, it can facilitate better collaboration and communication between a variety of technologies. The collaboration in question can be advantageous for numerous application areas, including smart traffic, smart grids, smart buildings, smart ecological settings, and intelligent logistics management. Big data management has long been the focus of numerous studies on the topic, leading to surveys on big data analytics. However, this research has always focused on IoT big data within the framework of massively parallel data analytics.

Some of the common platforms [5] offering big data processing and analytics are as follows:

a. *Apache Hadoop*: This open-source technology is used to store and handle massive volumes of data. It consists of a range of open-source software tools that help devices connected to a network of different computers to facilitate and solve issues with enormous amounts of unstructured data using the MapReduce programming style for processing. The Apache Hadoop software library is a framework that makes use of straightforward programming principles to enable the distributed processing of enormous data collections across multiple computer clusters. From single servers to thousands of workstations, each providing local processing and storage, it is designed to be flexible.

b. *1010data*: This platform deals in semi-structured data generated by IoT devices. 1010data is a platform that integrates analytics and data management with instant datasets and also allows people to share data among various companies. It offers a unique and uncommon cloud-based platform for sharing and discovering large amounts of data. An organization's service data warehouse is capable of providing quick response queries to various datasets that hold trillions of rows providing results through a Web browser in a familiar format such as Microsoft Excel. 1010data users can easily integrate and facilitate companies' data with various data from different sources like smart devices, traffic signals, employment data, and global positioning systems (GPS), enabling deeper business insights and more effective operations.

c. *Cloudera Data Hub*: Cloudera introduced the Enterprise Data Hub, designed solely to process data from IoT devices. It is an incredibly powerful and versatile cloud service on the Cloudera Data Platform that enables the development of contemporary, mission-driven, and data-driven applications with enterprise security, governance, scale, and control in a comfortable, quick, simple, and safe manner.

d. *SAP-Hana*: It consists of an inbuilt memory platform with the single aim of performing data analytics on data extracted from various IoT devices. It has increased performance in-memory databases, which can speed up real-time decisions and data-driven actions. SAP-HANA is a column-oriented, in-memory, relational database management system developed by SAP SE. Its major function as a database server is to retrieve stored data when requested by applications. The security concerns raised by DDoS attacks and the information provided by big data analytics lay the foundation of our approach for thwarting DDoS attacks in IoT networks.

3.3 BIG DATA ANALYTICS

A very crucial thing nowadays for any organization is generating big data analytics which contains various procedures including searching the databases, data mining, and data analyzing which is completely and solely used for enhancing the performance of the organizations. It is a vast process of scrutiny of huge datasets consisting of various datasets used to identify hidden correlations, unseen patterns, market trends, requirements of customers, and various types of information useful for business. This mechanism to analyze and verify huge amounts of sorted and unsorted data is useful for organizations required to enhance and expand businesses and plan future goals. Therefore, we can conclude that the main goal of big data analytics, as shown in Figure 3.1, is to support business associations by improving their comprehension of data so they can make the best decisions for their organizations.

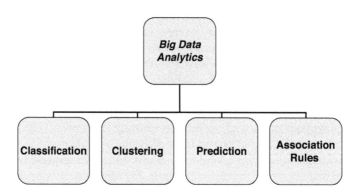

Figure 3.1 Overview of big data analytics tools.

Authors in [6] have stated that various expert data scientists and data miners are not able to extract useful information through traditional data extraction methods. However, this is now possible using big data analytic tools. These tools employ cutting-edge technology and techniques that enable the transformation of enormous amounts of structured, semi-structured, and unstructured data into polished, comprehensible metadata that is helpful for decision-making in any kind of business. They rely on algorithms to find patterns, trends, and correlations in the data over a range of time frames. After data analysis, these tools help visualize the findings and present them in various formats, such as graphs, pie charts, bar charts, and spatial charts, which are useful for an organization's decision-making process. Therefore, big data faces huge challenges and issues due to data complexity and scalability, underlying every algorithm designed to create patterns and correlations between various data. The algorithms generated should be scalable, and the techniques used for big data analytics should be able to generate timely and efficient results, leading to effective decision-making for any company.

So large amounts of infrastructure and surplus applications are very essential to aid in data parallelism. Apart from that, data sources like high-speed data streams, which are delivered through various sources and have varied formats, can make integration of a variety of sources for analytics solutions critical. Because the performance of the algorithms used in big data analysis today does not scale linearly with the increasing rise in processing resources, the challenge [7] stays focused on this. Big data analytics procedures take a long time to complete, and only a small number of tools can handle large amounts of datasets in a reasonable period of processing time. This results in feedback and instructions for users being provided. However, the bulk of the remaining tools rely on a convoluted trial-and-error approach to handle large volumes of heterogeneous data.

3.4 RELATION BETWEEN BIG DATA ANALYTICS AND IOT

Big data is created by the mesh network of IoT devices, so it is important to protect it from unscrupulous users to preserve its integrity. The use of analytics in big data is increasing, and big data analytics frameworks are becoming more popular. To address the difficulties associated with storing and processing vast amounts of data, writers in [8] suggested an IoT big data analytics framework. Essentially, big data and analytics processing tools are available and helpful for handling the massive amounts of data created by IoT devices. Among the notable mentions are SAP-Hana, 1010data, Cloudera Data Hub, Apache Hadoop, and others. Apache Hadoop and its ecosystem have attracted a lot of attention because the reliable ways of attack detection that have been used thus far have proven ineffective due to attackers changing their tactics. Its unique qualities are fault tolerance, scalability, and simplicity. Botcloud is a peer-to-peer MapReduce-based detection system that is scalable. The PageRank algorithm is at the center of the underlying idea. Hashdoop is a MapReduce framework for network anomaly detection that was proposed by Johan et al. To combat flooding attacks, the author in [9] suggested HADEC, a Hadoop-based Live DDoS Detection platform.

They have suggested leveraging the idea of window size entropy to identify DDoS attacks in Software Defined Networking (SDNs). The investigation was able to identify an attack, but no countermeasures were suggested. DDoS assaults threaten the abundance of knowledge contained in big data, with wormholes and blackholes being the main culprits. Organizations are shifting their focus from protecting big data to protecting it in an effort to remain ahead of the competition. The amount of work being done in this area shows how big data analytics can save the day. The method that is suggested uses data analytics to find potential black holes and wormholes and can help stop DDoS attacks.

Large, complex datasets with a range of data kinds can be analyzed and examined using big data analytics techniques to uncover previously unnoticed patterns, worldwide market trends, correlations between data, customs preferences, and crucial commercial information. By evaluating the data produced by big data analytics, the company may make wise judgments. According to [10], the main goal of big data analytics is to help business organizations better comprehend their data so they can make reliable and well-informed decisions. Data scientists can examine vast amounts of data with the use of big data analytics which may not be possible with a typical technique.

Large volumes of unstructured, semi-structured, and structured data can be transformed into useful findings and metadata for the analytical process using tools and techniques used in big data analytics. These analytical tools include algorithms that aid in the analysis of data patterns, correlations, and trends throughout a range of time periods. These analytical tools assist in visualizing the results of the analysis of the data in the form of graphs, tables, and charts, which are used for effective and strategic decision-making. Big data analysis can occasionally become quite difficult for several applications due to the complexity of the data and the scalability of algorithms that underpin these procedures.

Big data analytics is increasingly becoming a key factor in data provided by IoT devices, enabling enterprises to make smarter decisions. One of the main characteristics of IoT is its processing of data about devices that are connected. IoT big data analytics necessitates the immediate processing of enormous amounts of raw data and the storing of that data across a variety of storage methods. The procedure entails gathering unstructured and semi-structured data directly from the Web, where big data solutions start executing quick analytics with sophisticated queries to enable businesses to quickly analyze data, make decisions, and communicate with customers and other devices. Additionally, the interconnection between actual devices offers a way to share pertinent and helpful data across media through an integrated architecture and a shared operating system, facilitating the development of innovative applications.

Big data in IoT applications is a truly fascinating and amazing field right now. Both technologies are well-known in the IT and networking industries. Big data is not progressing as quickly as it should, yet these technologies are heavily interdependent and interwoven. The proliferation and acceptance of IoT have led to a greater amount of data integration, which has created opportunities for big data analytics development and applications. Nonetheless, the use of big data and IoT devices has improved IoT research and commercial models.

To begin with, any IoT data source can be linked to sensor devices and used with different apps to communicate with one another. Many different devices, such as smart traffic lights, cameras, and smart home appliances, communicate with one another and provide enormous amounts of data in various formats. This data, produced by multiple devices, is stored in the cloud using relatively affordable commodity storage mediums. Secondly, the enormous volume of data produced—based on the three Vs of volume, variety, and velocity—is referred to as big data. Big data files are stored in distributed fault-tolerant databases to accommodate this vast amount of data. Lastly, massive IoT datasets can be analyzed using analytics tools like Splunk, Spark, MapReduce, and Skytree. Figure 3.2 illustrates the four stages of data analytics, beginning with training data, moving to analytics tools, and eventually generating queries and reports.

IoT devices collect data from a variety of data sources, including smartphones, electronic devices, and smart devices [11]. Big IoT data [12] is generated in enormous amounts, at a very high speed, and primarily in an unstructured manner in Figure 3.1. Classifying the data comes first, followed by pattern recognition and mining, and last, prediction making [13].

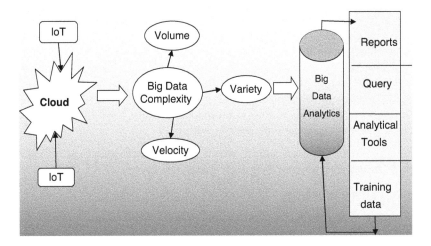

Figure 3.2 Relation between IoT and big data analytics.

For example, real-time traffic picture data can be generated in smart cities employing real-time traffic imagery, including traffic sensors, satellite shots, major traffic congestion, and traffic surveillance. Thus, traffic picture data from various sources is first grouped and then processed for data analysis to examine the real-time speed of the cars. Stream traffic sensors and road surveillance image data, on the other hand, are grouped as a single IoT data input, although having distinct data structures.

3.4.1 IoT architecture for big data analytics

There are numerous definitions of the IoT architectural idea, which are based on data abstraction and IoT domain identification. They offer reference models that show the connections between numerous IoT domains, including the medical industry, smart traffic, smart homes, smart devices, smart transportation, and smart health. Options for data abstraction and big data analytics design are provided by this architectural layout. In addition, it offers a reference architecture that expands upon the reference model. The authors of [14] have provided an end-to-end model for interaction among different stakeholders inside a cloud-centric IoT framework, as well as an IoT architecture that is integrated with cloud computing from within the center. Information representation and data analytics were used in the design of this architecture. However, the current architecture ignores the communication component in favor of IoT. The big data analytics and IoT architecture are shown in Figure 3.3. Figure 3.3 illustrates how a wireless network can link all the sensor devices and items related to the sensor layer. There are several options for this wireless network connection, including ZigBee, Bluetooth, Wi-Fi, and ultra-wideband. Web and internet connectivity is possible through the IoT gateway. The top layer is dedicated to big data analytics, where massive amounts of data are collected from sensor devices, stored on the cloud, and accessible via big data analytics apps. The dashboard and API administration features of these data analytic applications facilitate communication with the processing engine. In the context of holistic digital enterprise architecture, this notion is semi-automatically integrated. The main objective is to offer ample and robust decision support for complex and strategic business, IT environment architectural development, and effective assessment systems. As a result, choices pertaining to architecture and IoT are closely linked with the installation of code that enables customers to witness how an organization's architecture management collaborates with IoT.

3.4.2 Applications of Big IoT data analytics

3.4.2.1 Smart metering

In the IoT application depicted in the paper [15], various types of powerful machines, like smart grids, silos stock, water pumps, and temperature regulators produce huge amounts of data. Processing and generating

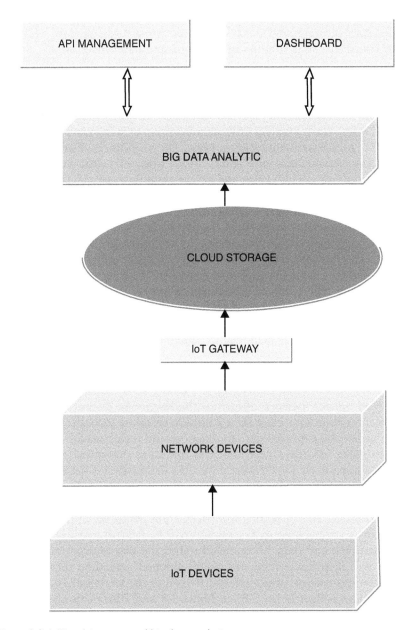

Figure 3.3 IoT architecture and big data analytics.

this data takes a lot of time but is useful for weather forecasting, temperature monitoring, and recording the consumption of electricity and water. Collecting and analyzing this data [16] is crucial for making decisions that can control crises, natural disasters, and the usage of natural resources.

Therefore, organizations should be capable and strong enough to generate data, extract the data, and make effective decisions using big data analytics, which transform huge volumes of data into actionable insights.

3.4.2.2 Smart transportation

This is the most demanding IoT system and the foundation of the smart city concept. It frequently combines the administration of smart cities with sophisticated and robust communication systems. The government of India has plans to transform several cities into "smart cities," where all traffic will be tracked and managed using RFID chips and license plates in every car. This is a useful way to monitor, identify, and track every vehicle. When the IoT is applied to every vehicle, it will help control traffic, deter auto theft, and potentially improve infrastructure design. With the use of sensors and satellite navigation, this technology may be used to improve the routing of all cars. A vast amount of data, including information about accidents, traffic jams, road conditions, and meteorological conditions, can be generated. Without the need for human intervention, it will even enable communication between two vehicles. With this system, all train and aviation traffic can be tracked. These vehicles also have sensors [17] that can offer real-time data for predicting errors, assessing whether equipment needs maintenance, and measuring engine health.

3.4.2.3 Smart supply chains

Embedded sensor devices allow for bidirectional communication and remote accessibility for more than 2 million elevators globally. Off-site personnel utilize the captured data to diagnose and repair machines, enabling them to make informed decisions that minimize downtime and improve customer happiness. Furthermore, massive IoT analytics helps the supply chain make informed decisions. It also plays a crucial role in providing future scope as technologies like sensors, cloud, GPS, and RFID [18] can help in tracking and providing information on goods being transferred. It will act as a strong support system and will be fruitful to all retailers and manufacturers; the data generated through the communication of these devices will help track items and goods shipped, including their position and other information until each is delivered to the final end. The different types of data collected via GPS technologies and RFID can allow supply chain managers to have accurate delivery information and improve automated shipment by predicting the time of arrival [19].

3.4.2.4 Smart agriculture

Smart agriculture is a big boon these days. As a sensor, GPS devices can be installed in fields to keep track of information on plants and their growth, help in keeping track of weather conditions, and temperature monitoring can also be done through these devices [20]. This is beneficial for farmers as it helps them keep good track of their fields and make strategic decisions after results analysis and monitoring. It depicts the harvesting conditions and moisture in the soil and helps in timely and controlled irrigation [21].

3.4.2.5 E-commerce

Huge IoT data analytics offers excellent tools and technologies to process huge data in real time and get quick, reliable findings that may be used to make wise decisions [22,23]. Big IoT data demonstrates real-time data processing capabilities and data heterogeneity. Building a suitable smart environment has new benefits and challenges as a result of big data and IoT integration. Big IoT data analytics is widely used in nearly every industry. But the most successful areas of using these big data analytics are in revenue growth, e-commerce, analysis of customer records, tracking and expanding customer relations, predictions of sale forecasts, risk management, customer segmentation, and product optimization.

3.4.2.6 Smart cities

Smart cities [24,25] are upcoming opportunities where big data generated by smart devices can lead to effective decisions that enhance the economy of cities and the overall growth of the country. The smart city concept can be integrated into healthcare systems, smart transportation, and e-commerce. The combined powers of big data and the IoT can revolutionize every economic sector by processing, streaming, and manipulating data in real time. Monitoring various aspects of data and casting data is also useful for predicting natural disasters or unknown anomalies [26].

3.4.2.7 Healthcare

Recently we have seen a huge amount of growth in the health monitoring system. Smart devices generate a tremendous amount of data which is useful for monitoring patients' data and keeping records of all patients. The data generated by smart devices can also be useful in disease diagnosis, and future treatments can be predicted through the analytic data generated by these devices. Doctors can review the huge data generated by these smart systems and provide effective treatment to patients. We can also apply data analytics to temperature monitors and blood glucose levels, and analyze the health of patients. All these advancements can ease the work for doctors, and the health of patients can be reviewed accurately at regular intervals (Table 3.1).

Table 3.1 Comparison of various applications of IoT big data analytics

Area of application	Benefits	IoT device	Data source
Smart metering	Prediction of resources like water and electricity	RFID devices Sensors Camera equipment	Text, images
Smart cities	Prediction of economic growth and development	Sensors RFID Cameras GPS devices	Text, video, audio, images
Smart healthcare	Patient record keeping and analysis	Sensor devices Camera devices Smart monitors	Text, video, audio, images
Smart transportation	Improve traffic monitoring and reducing congestion	Sensor devices RFID devices Camera equipment GPS devices	Text, video, audio, images
E-commerce	Allow smooth process of goods transportation	Sensor devices RFID devices Camera equipment Mobile devices	Text, videos, images
Smart agriculture	Monitoring of weather conditions and soil condition	Sensor devices RFID devices Camera equipment	Text, videos, images, audio
Smart supply chain	Helps in strategic decision-making	Sensor devices RFID devices GPS devices	Text, videos, images, audio

3.5 CONCLUSION

Big data and the IoT have a close relationship that is transforming data generation, processing, and utilization through their synergistic partnership. Because of its vast network of linked devices, IoT continuously generates enormous amounts of data from sensors, actuators, and other embedded technologies. This flood of various forms of data, including photographs, sensor readings, and real-time information, requires the use of big data analytics. Big data provides the infrastructure and tools needed to manage the vast amount and diversity of data produced by IoT devices. Large datasets can be processed, stored, and analyzed, revealing important trends and insights that support well-informed decision-making. The seamless alignment of the real-time nature of IoT data with the real-time or near-real-time processing capabilities of big data analytics enables quick reactions to dynamic events. As both ecosystems grow, their combination tackles scalability issues while facilitating predictive modeling and decision support. Furthermore, the integration of IoT and big data emphasizes the importance of crucial security and privacy protocols to protect sensitive data created by networked devices. Collectively, they open up new avenues and revolutionize the field of data-driven insights by paving the way for creative solutions across various sectors.

REFERENCES

1. Sun, Y., Song, H., Jara, A.J. and Bie, R., 2016. Internet of Things and big data analytics for smart and connected communities. *IEEE Access*, *4*, pp. 766–773.
2. Gohar, M., Ahmed, S.H., Khan, M., Guizani, N., Ahmed, A. and Rahman, A.U., 2018. A big data analytics architecture for the Internet of Small Things. *IEEE Communications Magazine*, *56*(2), pp. 128–133.
3. Yacchirema, D.C., Sarabia-Jácome, D., Palau, C.E. and Esteve, M., 2018. A smart system for sleep monitoring by integrating IoT with big data analytics. *IEEE Access*, *6*, pp. 35988–36001.
4. Ur Rehman, M.H., Ahmed, E., Yaqoob, I., Hashem, I.A.T., Imran, M. and Ahmad, S., 2018. Big data analytics in industrial IoT using a concentric computing model. *IEEE Communications Magazine*, *56*(2), pp. 37–43.
5. Ahmad, S., Jha, S., Abdeljaber, H.A.M., Rahmani, M.K.I., Waris, M.M., Singh, A. and Yaseen, M., 2022. An Integration of IoT, IoC, and IoE towards building a green society. *Scientific Programming*, *2022*, p. 2673753.
6. Fawzy, D., Moussa, S.M. and Badr, N.L., 2022. The internet of things and architectures of big data analytics: Challenges of intersection at different domains. *IEEE Access*, *10*, pp. 4969–4992.
7. Ahmed, E., Yaqoob, I., Hashem, I.A.T., Khan, I., Ahmed, A.I.A., Imran, M. and Vasilakos, A.V., 2017. The role of big data analytics in Internet of Things. *Computer Networks*, *129*, pp. 459–471.
8. Marjani, M., Nasaruddin, F., Gani, A., Karim, A., Hashem, I.A.T., Siddiqa, A. and Yaqoob, I., 2017. Big IoT data analytics: Architecture, opportunities, and open research challenges. *IEEE Access*, *5*, pp. 5247–5261.
9. Paul, A., Ahmad, A., Rathore, M.M. and Jabbar, S., 2016. Smartbuddy: Defining human behaviors using big data analytics in social internet of things. *IEEE Wireless Communications*, *23*(5), pp. 68–74.
10. Arya, A. and Jha, S., 2021. "An analytical review on data privacy and anonymity in "Internet of Things (IoT) enabled services." In *2021 3rd International Conference on Advances in Computing, Communication Control and Networking (ICAC3N)*, pp. 1432–1438. IEEE, Piscataway, NJ.
11. Martis, R.J., Gurupur, V.P., Lin, H., Islam, A. and Fernandes, S.L., 2018. Recent advances in big data analytics, internet of things and machine learning. *Future Generation Computer Systems*, *88*, pp. 696–698.
12. Dey, N., Hassanien, A.E., Bhatt, C., Ashour, A. and Satapathy, S.C. eds., 2018. *Internet of Things and Big Data Analytics toward Next-Generation Intelligence* (Vol. 35). Berlin: Springer.
13. Hameed, S., Khan, F.I. and Hameed, B., 2019. Understanding security requirements and challenges in Internet of Things (IoT): A review. *Journal of Computer Networks and Communications*, *2019*, pp. 648–651.

14. Lee, J., Ardakani, H.D., Yang, S. and Bagheri, B., 2015. Industrial big data analytics and cyber-physical systems for future maintenance & service innovation. *Procedia CIRP*, *38*, pp. 3–7.
15. Lee, C.K.M., Yeung, C.L. and Cheng, M.N., 2015. "Research on IoT based cyber physical system for industrial big data analytics." In *2015 IEEE International Conference on Industrial Engineering and Engineering Management (IEEM)*, pp. 1855–1859. IEEE, Piscataway, NJ.
16. Rizwan, P., Suresh, K. and Babu, M.R., 2016. "Real-time smart traffic management system for smart cities by using Internet of Things and big data." In *2016 International Conference on Emerging Technological Trends (ICETT)*, pp. 1–7. IEEE, Piscataway, NJ.
17. Yaqoob, I., Ahmed, E., Hashem, I.A.T., Ahmed, A.I.A., Gani, A., Imran, M. and Guizani, M., 2017. Internet of things architecture: Recent advances, taxonomy, requirements, and open challenges. *IEEE Wireless Communications*, *24*(3), pp. 10–16.
18. Li, L., Xu, D. and Zhao, S., 2014. "The Internet of Things: A survey," In *Information Systems Frontiers*, pp. 243–299. New York: Springer. https://doi.org/10.1007/s10796-014-9492-7
19. Ahmad, S., Jha, S., Alam, A., Alharbi, M. and Nazeer, J., 2022. Analysis of intrusion detection approaches for network traffic anomalies with comparative analysis on botnets (2008–2020). *Security and Communication Networks*, *2022*(4), pp. 1–11.
20. Al-Sai, Z.A., Husin, M.H., Syed-Mohamad, S.M., Abdin, R.M.D.S., Damer, N., Abualigah, L. and Gandomi, A.H., 2022. Explore big data analytics applications and opportunities: A review. *Big Data and Cognitive Computing*, *6*(4), p. 157.
21. Oesterreich, T.D., Anton, E., Teuteberg, F. and Dwivedi, Y.K., 2022. The role of the social and technical factors in creating business value from big data analytics: A meta-analysis. *Journal of Business Research*, *153*, pp. 128–149.
22. Sasaki, Y., 2021. A survey on IoT big data analytic systems: Current and future. *IEEE Internet of Things Journal*, *9*(2), pp. 1024–1036.
23. Shi, Y. and Shi, Y., 2022. Big data and big data analytics. *Advances in Big Data Analytics: Theory, Algorithms and Practices*, *2022*, pp. 3–21.
24. Jha, S., Jha, N., Prashar, D., Ahmad, S., Alouffi, B. and Alharbi, A., 2022. Integrated IoT-based secure and efficient key management framework using hashgraphs for autonomous vehicles to ensure road safety. *Sensors*, *22*(7), p. 2529.
25. Oesterreich, T.D., Anton, E., Teuteberg, F. and Dwivedi, Y.K., 2022. The role of the social and technical factors in creating business value from big data analytics: A meta-analysis. *Journal of Business Research*, *153*, pp. 128–149.
26. Singh, K. and Jha, S., 2021. "Cyber threat analysis and prediction using machine learning." In *2021 3rd International Conference on Advances in Computing, Communication Control and Networking (ICAC3N)*, pp. 1981–1985. IEEE, Piscataway, NJ.

Chapter 4

Data wrangling in smart devices

Sudan Jha, Sarbagya Ratna Shakya, and Sultan Ahmad

4.1 INTRODUCTION

4.1.1 Background

The Internet of Things (IoT) has burgeoned into a technological phenomenon, reshaping the way we interact with our surroundings. The proliferation of interconnected devices has reached unprecedented levels, embedding intelligence into everyday objects, and creating a seamless web of data exchange. This transformative growth has not only revolutionized industries but also generated an overwhelming deluge of data.

Overview of IoT Growth: The exponential rise of IoT is evident in the myriad of devices that now permeate our homes, cities, and industries. From smart thermostats to connected vehicles, the IoT landscape is expanding, promising unprecedented convenience and efficiency. The integration of sensors and communication technologies has given rise to a vast network of interconnected devices, collectively contributing to the fabric of IoT (Figure 4.1).

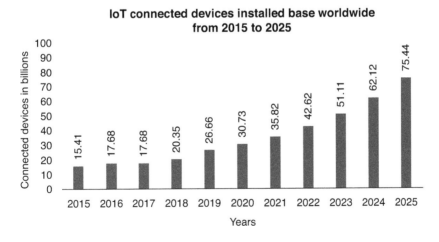

Figure 4.1 IoT-connected devices from 2015 to 2025 [1].

DOI: 10.1201/9781003530572-4

Data Proliferation: However, this surge in connected devices comes with a significant consequence—an unprecedented proliferation of data. Every sensor reading, user interaction, and environmental variable contributes to an immense volume and variety of data points. As these devices communicate and share information in real time, the sheer scale of data generated poses both a challenge and an opportunity.

4.1.2 Purpose of data wrangling in IoT

Data's Role in IoT: At the heart of IoT lies the pivotal role of data. It serves as the lifeblood, fueling the operations and insights that make IoT applications thrive. The effectiveness of IoT ecosystems is intrinsically tied to the quality, relevance, and accessibility of the data they generate. From monitoring environmental conditions to optimizing industrial processes, data acts as the conduit through which IoT applications fulfill their potential [2].

Importance of Data Wrangling: However, the raw data generated by IoT devices is often a chaotic symphony of diverse formats, structures, and levels of precision. This is where the concept of data wrangling steps in [3]. Data wrangling involves the art and science of cleaning, structuring, and enriching raw data to make it usable for analysis. In the realm of IoT, where data arrives in varied forms and dimensions, effective data wrangling becomes indispensable. It is the key to unlocking meaningful insights, enabling informed decision-making, and harnessing the full potential of the data deluge [4].

Imagine you are a chef preparing a feast for a grand celebration. Your kitchen is filled with a variety of ingredients—fresh vegetables, succulent meats, and a plethora of spices. Now, each ingredient needs to be chopped, sliced, and prepared in a specific way to bring out its best flavors. This meticulous preparation is akin to what data wrangling does for the vast amount of information collected in the world of technology.

In the realm of data, we have a similar scenario. Data is generated from various sources, just like ingredients in a kitchen. This could be anything from your shopping preferences online, the steps you take with your fitness tracker, to the temperature readings from smart thermostats in homes. However, this data often comes in different formats, is messy, and needs to be processed before it can be used effectively.

Here's where data wrangling steps in. It's like having a master chef in your kitchen who takes all those diverse ingredients, cleans them up, chops them into the right sizes, and makes sure they blend seamlessly to create a delightful dish. Similarly, data wrangling takes messy, raw data and transforms it into a clean, organized, and meaningful form.

Real-Time Example: *Let's say you work for a delivery company, and you're tasked with optimizing delivery routes for your drivers. You have a ton of data coming in—addresses, delivery times, traffic conditions, and more. This data might be scattered and unorganized, making it challenging to derive useful insights.*

Data wrangling in this context would involve cleaning up this data, ensuring all addresses are formatted correctly, removing any duplicate information, and filling in any missing details. It's about preparing this data in a way that makes it easy to analyze.

Once the data is wrangled, you can now use it to figure out the most efficient routes for your delivery drivers. This could mean they spend less time on the road, deliver packages faster, and maybe even save on fuel costs.

In essence, data wrangling is the behind-the-scenes work that makes the data useful, just like the chef's preparation that turns raw ingredients into a delicious meal. It ensures that the information gathered is ready to be used for making informed decisions and improvements in various areas, making businesses run more smoothly and effectively.

4.1.3 Significance of efficient data handling

Enhancing System Performance: Efficient data handling, orchestrated through the meticulous process of data wrangling, plays a pivotal role in elevating the overall performance of IoT systems. The agility with which data is processed and transformed directly influences the responsiveness of applications. Faster data processing not only enhances the real-time capabilities of IoT solutions but also ensures that critical insights are delivered promptly, contributing to improved operational efficiency [5].

Data Quality for Analytics: Beyond speed, the quality of the data is paramount, especially when it comes to analytics. Clean, accurate, and well-structured data forms the bedrock of meaningful analytics in the IoT context. The reliability of insights derived from analytics hinges on the integrity of the input data. Effective data wrangling ensures that the data fed into analytical models is of the highest quality, empowering organizations to derive actionable insights and make informed decisions.

In conclusion, the growth of IoT, coupled with the surge in data, necessitates a strategic approach to data handling. Data wrangling emerges as the linchpin in this narrative, offering a pathway to harness the vast potential embedded in the interconnected web of devices. As we navigate the intricate landscape of IoT, the efficiency and effectiveness of data wrangling will continue to shape the trajectory of innovation and progress.

4.2 OVERVIEW OF SMART DEVICES IN IOT

IoT is not just a technological evolution; it's a revolution that revolves around the seamless integration of smart devices. These devices, embedded with sensors, processors, and communication modules, form the building blocks of the interconnected world we inhabit today. As we delve into the landscape of IoT, understanding the essence and role of smart devices becomes paramount.

4.2.1 Definition and types of smart devices

Definition: Smart devices, in the context of IoT, refer to everyday objects enhanced with embedded technologies that enable them to gather, process, and exchange data. These devices, ranging from household appliances to industrial machinery, are equipped with sensors that facilitate communication and interaction, both with users and with other connected devices [6]. The 'smart' prefix denotes their ability to operate autonomously or in response to user input, often leveraging artificial intelligence for enhanced functionality.

4.2.1.1 Types of smart devices

The spectrum of smart devices in IoT is diverse and continually expanding. Home automation has witnessed the proliferation of devices like smart thermostats, connected refrigerators, and intelligent lighting systems. Wearables, such as fitness trackers and smartwatches, seamlessly integrate with our daily lives, monitoring health metrics [7] and providing personalized insights. In industrial settings, IoT-enabled sensors enhance the efficiency of machinery, contributing to the paradigm of Industry 4.0. Understanding the nuances of these devices is essential for grasping the breadth of IoT applications.

4.2.2 Role of smart devices in the IoT ecosystem

At the core of the IoT ecosystem are these intelligent nodes—the smart devices that collectively form a dynamic and interconnected network. Their role goes beyond mere automation [8]; they serve as the conduits through which data flows, creating a symbiotic relationship between the physical and digital realms. Smart devices act as the eyes, ears, and sometimes even the limbs of the IoT infrastructure, facilitating the exchange of information that powers the entire system.

In a smart home scenario, for instance, sensors embedded in devices like smart thermostats and security cameras collect data on temperature, occupancy, and environmental conditions. This data is then transmitted to a central hub or cloud, where it undergoes analysis and triggers actions. The interconnectedness of these devices results in a cohesive ecosystem where information is not isolated but shared, allowing for more comprehensive insights and responsive functionalities.

4.2.3 Importance of data generated by smart devices

The true essence of smart devices in the IoT narrative lies in the data they generate. Every interaction, every sensor reading, contributes to a continuous stream of data that serves as the lifeblood of IoT applications. The importance of this data cannot be overstated [9].

Fueling Decision-Making: The data generated by smart devices becomes the foundation for informed decision-making. In industrial settings, machinery equipped with sensors provides real-time performance metrics, enabling predictive maintenance and minimizing downtime. In healthcare [10], wearable devices monitor vital signs, offering a proactive approach to health management. The actionable insights derived from smart device data empower individuals and organizations alike to make decisions that are not just reactive but proactive.

Enabling Personalization: In the realm of consumer-oriented IoT applications, the data generated by smart devices is instrumental in personalization. Smart thermostats learn user preferences over time, optimizing temperature settings for comfort and energy efficiency. Streaming platforms use viewing history data to recommend personalized content. The ability to tailor experiences based on individual behavior is a testament to the transformative power of smart device-generated data.

As we navigate the landscape of smart devices in IoT, it becomes evident that these devices are not merely gadgets with enhanced capabilities; they are catalysts for a data-driven revolution. Understanding their types, roles, and the significance of the data they generate sets the stage for comprehending the intricate dynamics of the IoT ecosystem. In the chapters that follow, we'll delve deeper into the mechanisms that enable these devices to communicate, the challenges they pose, and the solutions that drive the evolution of IoT.

4.3 DATA GENERATION IN SMART DEVICES

At the heart of IoT lies the relentless production of data by smart devices. So, we need to delve into the intricate mechanisms governing this data generation process, exploring the role of sensors, the myriad types of data produced, and the staggering volumes and velocities at which this information flows.

4.3.1 Sensors and data sources

Sensory Symphony: Smart devices owe their intelligence to an array of sensors that transform the physical world into a realm of data. These sensors act as the sensory organs of the IoT ecosystem, perceiving changes in the environment and converting them into a language understandable by machines. Common sensors include accelerometers, gyroscopes, temperature sensors, cameras, and environmental sensors like humidity and pressure detectors.

Data Sources: The diversity of sensors translates into a multitude of data sources. In healthcare devices, biosensors capture vital signs, while in smart homes, motion detectors and light sensors contribute to the understanding of user behavior. Industrial IoT relies on sensors embedded in machinery to monitor performance and detect anomalies. Each sensor serves a specific purpose, collectively creating a rich tapestry of data that characterizes the IoT landscape.

4.3.2 Types of data produced by smart devices

Structured and Unstructured Symphony: Smart devices generate a spectrum of data types, ranging from structured to unstructured. Structured data adheres to a predefined format, often tabular, and is highly organized. This includes sensor readings, timestamps, and categorical data. Unstructured data, on the other hand, lacks a predefined structure and encompasses textual data, images, and videos [11]. The amalgamation of these data types provides a comprehensive view of the device's surroundings and interactions.

Examples of Data Types: Health monitoring devices produce structured data such as heart rate readings and step counts, while a smart security camera generates unstructured data in the form of video feeds. Understanding the nuances of these data types is essential for extracting meaningful insights and implementing effective data-wrangling strategies.

4.3.3 Volume and velocity of data generation

Inundation of Data: The proliferation of smart devices contributes to an unprecedented volume of data. The sheer number of devices, each continuously producing streams of data, results in a data deluge that organizations must navigate. This volume poses both a challenge and an opportunity, requiring robust infrastructures and analytical frameworks to harness the potential insights within [12].

Velocity of Data Flow: The velocity at which data flows from smart devices is equally crucial. Real-time applications, such as autonomous vehicles or industrial control systems, demand instantaneous processing of data to enable swift decision-making. The velocity of data generation is intertwined with the responsiveness of IoT applications, making it imperative to design systems that can handle the dynamic nature of incoming data.

As we unravel the intricacies of data generation in smart devices, it becomes evident that the synergy of sensors, the diversity of data types, and the magnitude of data volume and velocity shape the landscape of IoT. In the subsequent chapters, we will explore the journey of this data from generation to utilization, emphasizing the pivotal role it plays in driving the innovation and transformative power of IoT applications.

4.4 CHALLENGES IN SMART DEVICE DATA

The vast landscape of smart devices brings forth a multitude of challenges in handling their generated data [13]. This chapter meticulously explores these challenges, unveiling the intricate tapestry of variety, complexity, data quality concerns, and the ever-looming specters of security and privacy [14].

4.4.1 Variety and complexity of data

Data: A Diverse Tapestry: Smart devices, each with its unique purpose and functionality, contribute to the diverse tapestry of data. From structured numerical readings to unstructured multimedia files, the variety poses a significant challenge. The complexity arises not only from the types of data but also from the interplay of these types within the IoT ecosystem. Understanding and managing this diversity is fundamental for effective data wrangling.

Interconnectedness of Data Types: Structured data, such as sensor readings, often needs to be correlated with unstructured data like images or textual information for a holistic understanding. This interconnectedness adds an extra layer of complexity, requiring sophisticated approaches to ensure that disparate data types seamlessly integrate.

4.4.2 Data quality and integrity issues

The Imperative of Quality: The quality of data holds paramount importance in the realm of smart devices. Erroneous readings, inaccuracies, or missing data can significantly impact decision-making processes. Ensuring data quality involves addressing issues like inaccuracies in sensor measurements, calibration errors, and the need for real-time validation mechanisms.

Integrity Challenges: Maintaining the integrity of data throughout its lifecycle is a formidable task. The journey from sensor to storage introduces vulnerabilities, and data may be susceptible to corruption, tampering, or loss. Strategies for data integrity assurance become imperative to uphold the reliability and trustworthiness of the information being generated.

4.4.3 Security and privacy concerns

Sentinels at the Gate: As conduits of sensitive information, smart devices become prime targets for malicious actors. Security concerns encompass a spectrum of threats, from unauthorized access to the manipulation of data streams. Protecting the confidentiality, integrity, and availability of data demands robust security measures, including encryption, authentication, and secure communication protocols.

Privacy: A Delicate Balancing Act: The very nature of smart devices involves constant data collection, raising profound privacy concerns. Balancing the benefits of data-driven insights with the protection of individual privacy becomes a delicate dance. Striking the right balance involves implementing privacy-preserving techniques, transparent data usage policies, and adhering to regulatory frameworks.

As we navigate the labyrinth of challenges in smart device data, it becomes evident that effective solutions necessitate a comprehensive understanding of the intricacies involved. Subsequent chapters will delve into strategies and innovations aimed at overcoming these challenges, ensuring that the wealth of data generated by smart devices becomes a powerful catalyst for progress rather than a stumbling block.

4.5 DATA WRANGLING TECHNIQUES

In the intricate world of IoT, where data reigns supreme, mastering the art of data wrangling is pivotal. This chapter embarks on a journey through the nuanced techniques that empower data scientists and analysts to transform raw IoT data into a refined, actionable resource. From the initial steps of preprocessing to the intricate dance of cleaning and handling inconsistencies, every facet of data wrangling unfolds.

4.5.1 Introduction to data wrangling

Unveiling the Craft: Data wrangling, often referred to as data munging, is the unsung hero of the data science realm. It encapsulates the set of processes involved in cleaning, structuring, and enriching raw data into a format suitable for analysis. In the context of IoT, where data streams from multifarious sources, the need for a systematic approach to wrangling becomes paramount.

4.5.2 Preprocessing steps for raw IoT data

The Significance of IoT: IoT data, by its very nature, is messy. It comes in different formats, exhibits inconsistencies, and may contain missing values. Effective data wrangling bridges the gap between the chaotic raw data and the structured datasets required for meaningful analysis. It is the linchpin in the data pipeline, ensuring that insights drawn from IoT ecosystems are accurate and reliable.

4.5.3 Cleaning and transforming data

Navigating the Cleaning Odyssey: Cleaning data is akin to a meticulous odyssey. In the realm of IoT, this process involves detecting and rectifying errors, handling outliers, and ensuring uniformity across diverse datasets. Cleaning is not merely about erasing imperfections; it's about preparing the data for meaningful analysis.

Transformation: The Alchemy of Data: Transformation unleashes the alchemy of data, converting it into a refined, enriched version. In the IoT landscape, this entails converting data types, aggregating information, and creating derived features that augment the analytical potential of datasets. The transformative phase lays the groundwork for extracting actionable insights.

4.5.4 Handling missing and inconsistent data

The Lacuna of Missing Data: Missing data is a pervasive challenge in IoT datasets. Addressing this lacuna involves strategic imputation methods, where missing values are estimated based on patterns, statistical measures, or predictive modeling. The goal is to fill the gaps judiciously, ensuring that the imputed values align with the overall characteristics of the dataset.

Confronting Inconsistencies: Inconsistencies, arising from discrepancies in data formats or units, demand a concerted effort. Standardization becomes the beacon, harmonizing disparate elements into a cohesive whole. The chapter explores techniques for detecting and reconciling inconsistencies, fostering a dataset that is not only complete but also harmonized for coherent analysis. As we delve into the heart of data-wrangling techniques, each facet unfolds as a critical step in the journey from raw data to actionable insights. The subsequent sections will further dissect these techniques, offering a detailed roadmap for practitioners navigating the complex terrain of IoT data wrangling.

4.6 INTEGRATION OF DATA FROM MULTIPLE DEVICES

In the intricate landscape of IoT, where a myriad of devices collaborate to generate torrents of data, the integration of this diverse data becomes a paramount endeavor. This chapter unravels the significance, challenges, and strategies associated with the integration of data from multiple devices. From the importance of harmonizing disparate data streams to navigating the complexities inherent in diverse devices, every facet of data integration is meticulously explored.

4.6.1 Importance of data integration

The Symphony of Data: In the orchestration of IoT ecosystems, where devices function as instrumental nodes, data integration emerges as the conductor orchestrating a harmonious symphony. The importance of weaving together data from various devices lies in the ability to create a holistic narrative. Integrated data provides a comprehensive view, enabling more accurate analytics, actionable insights, and informed decision-making.

Holistic Insights: Isolated data streams offer fragmented glimpses into the operations of individual devices. Through integration, these fragments coalesce into a panoramic view, revealing correlations, patterns, and anomalies that remain obscured in siloed datasets. This chapter delves into the transformative power of integrated data in fostering a holistic understanding of the IoT landscape.

4.6.2 Challenges in integrating data from diverse devices

Diverse Devices, Divergent Challenges: The diversity of devices within IoT ecosystems presents a kaleidoscope of challenges in the integration process. From variations in data formats and protocols to differences in temporal and spatial resolutions, each device introduces nuances that demand careful consideration. This chapter navigates through these challenges, shedding light on the intricacies of harmonizing data from devices with diverse characteristics.

Interoperability Conundrum: Achieving seamless interoperability between devices with distinct communication protocols and standards poses a significant hurdle. The chapter explores the complexities of interoperability, examining strategies to bridge the divide and facilitate smooth data integration. From middleware solutions to standardized communication frameworks, the arsenal against interoperability challenges is dissected.

4.6.3 Strategies for effective data integration

Architecting Integration: Effecting successful data integration necessitates a strategic architectural approach. This chapter unfolds various integration architectures, from hub-and-spoke models to decentralized approaches. Each architecture is scrutinized for its suitability in different IoT scenarios, providing a roadmap for architects and practitioners navigating the integration terrain.

Real-time Integration: In the dynamic realm of IoT, where real-time insights steer decision-making, the importance of real-time data integration cannot be overstated. This chapter explores strategies for achieving real-time integration, from stream processing to event-driven architectures. The goal is to enable organizations to harness the power of integrated data instantaneously.

4.7 DATA WRANGLING TOOLS AND TECHNOLOGIES

Data wrangling serves as the alchemical process that transforms raw, unruly data into refined insights [3]. This chapter embarks on an expedition through the landscape of data-wrangling tools and technologies, deciphering the array of options available, comparing leading platforms, and establishing selection criteria tailored to the unique challenges of IoT data [15].

4.7.1 Overview of available data wrangling tools

The Toolbox Unveiled: A multitude of tools stand ready to assist in the intricate dance of data wrangling. From open-source champions like Pandas and OpenRefine to commercial heavyweights such as Trifacta and Alteryx, this section provides a comprehensive overview of the diverse tools available. Each tool's strengths, limitations, and primary applications are dissected, offering readers a nuanced understanding of the options at their disposal.

Pandas: Pythonic Powerhouse: Pandas, a stalwart in the realm of data manipulation in Python, takes center stage. This chapter explores its capabilities in handling diverse data structures, executing operations, and seamlessly integrating with other data science libraries. A hands-on exploration of Pandas illuminates its utility in taming the intricacies of IoT data.

OpenRefine: Crafting Clean Data: OpenRefine emerges as a versatile tool for cleaning and transforming messy data. This chapter delves into its capabilities, showcasing its user-friendly interface and powerful operations. A step-by-step guide illustrates how OpenRefine can be wielded to wrangle data into submission, making it an indispensable ally in the data refinement process.

4.7.2 Comparison of popular data wrangling platforms

Trifacta: Orchestrating Data Alchemy: Trifacta, a leading data wrangling platform, steps into the spotlight. This section conducts a detailed comparison, pitting its capabilities against other platforms. The discussion covers Trifacta's intelligent features, collaborative functionalities, and its ability to handle the intricacies of IoT data. Real-world case studies underscore Trifacta's prowess in orchestrating data alchemy.

Alteryx: Bridging the Data Divide: Alteryx, known for its end-to-end analytics platform, undergoes scrutiny in this comparative analysis. This chapter unveils its capabilities in data blending, preparation, and analytics. Practical use cases elucidate how Alteryx navigates the complexities of diverse datasets, acting as a bridge between raw data and actionable insights.

4.7.3 Selection criteria for data wrangling tools in IoT

Tailoring Tools to IoT Challenges: Selecting the right data-wrangling tool for IoT demands a discerning eye. This section outlines specific criteria tailored to the challenges posed by IoT data. From scalability and real-time processing to compatibility with IoT protocols, the selection process is demystified. This chapter empowers readers with a decision-making framework, ensuring that the chosen tool aligns seamlessly with the unique demands of wrangling IoT data.

Scalability: Managing the Data Deluge: In the IoT realm, where data deluge is the norm, scalability emerges as a critical criterion. This chapter explores how data-wrangling tools cope with the challenges of handling voluminous and streaming IoT data. Case studies illustrate how scalability becomes a linchpin in ensuring the effectiveness of data-wrangling processes.

Real-time Processing: Navigating the Time Dimension: IoT operates in real-time, demanding tools that can keep pace. This section dissects the importance of real-time processing capabilities in data-wrangling tools. Practical scenarios showcase how real-time processing enhances the agility and responsiveness of data wrangling, unlocking timely insights for IoT applications.

Compatibility with IoT Protocols: Speaking IoT's Language: IoT devices communicate through specific protocols, and data-wrangling tools must speak their language. This chapter explores the importance of compatibility with prevalent IoT protocols. Use cases elucidate how tools adept in handling these protocols streamline the wrangling process, ensuring a seamless flow of data from devices to insights.

4.8 CASE STUDIES

With the use of multiple sensors, devices, and systems providing raw data in real time, this raw data needs to be changed into a format suitable for further processing. These processes, which involve the processing of data such as cleaning, structuring, and transforming, are known as data wrangling. Data wrangling is important for the accuracy, completeness, and effectiveness of data to generate information from the analysis of the data. These have been applied to some applications in the real world.

4.8.1 Real-world examples of data wrangling in IIoT

Many real-world companies have already applied data wrangling to improve efficiency, optimize profit, and lessen environmental impacts. Some examples include:

a. *UPS*: This shipping company uses data collected from its vehicles, such as speed, mileage, routes, and fuel consumption, to optimize its operations. These raw data collected from the sensors are used after data wrangling for analysis to save money and improve operational efficiency.

b. *Amazon*: Data wrangling is used in companies like Amazon to prepare data that generates insights to improve its products and services, such as personalization, recommendations, fraud detection, and customer satisfaction. Amazon also provides data wrangling tools like Amazon SageMaker Data Wrangler, which help access and select data from various sources, generate data insights and quality reports, transform data with over 300 built-in transformations, estimate model accuracy, and diagnose issues.

c. *Manufacturing Industry*: Data wrangling is used in the manufacturing industry to prepare data for optimizing the supply chain, asset management, and operational management. The data collected from various sources such as sensors, machines, and databases are used for accessing, selecting, transforming, validating, and publishing data.

This helps the manufacturing industry to automate and simplify data with more accurate data analysis.

4.8.2 Data wrangling applications

Data wrangling is applicable in various domains where data is collected from various sources and needs to be cleaned, processed, and prepared for analysis. This process is essential to ensure that the data is reliable, consistent, and of high quality, enabling it to be analyzed, modeled, and visualized. Some of the sectors where data wrangling can be applied include:

a. *Customer Analytics*: With data collected from customers regarding any product or services from sources like customer feedback, website interactions, customer behavior, and transaction records, data wrangling can be used for exploratory analysis, data cleaning, creating data structure, and data storage. This data can be transformed into a unified format and used for analyzing customer product feedback, preferences, and patterns, allowing for suitable changes in the manufacturing process to improve quality.

b. *Predictive Modeling*: With enormous amounts of data coming from different sources in the industry, data wrangling can prepare this data to be suitable for training machine learning models. The process involves handling missing values, normalizing, encoding categorical variables, and other data preprocessing techniques to increase the accuracy of predictive models.

c. *Market Research and Surveys*: Data collected through surveys and questionnaires are vital to understanding the market sentiments. The data from these sources needs to be clean, outliers should be removed, missing values need to be handled, and this unstructured data needs to be transformed into a more structured format such that it can be used to analyze customer preferences, market trends, and market insights.

d. *Financial Analysis*: Data wrangling can also be used to clean, remove errors, and perform calculations and aggregations of financial data of the industry or factories. This financial data collected from various sources needs to be organized such that it will provide accurate financial analysis with risk assessment for optimal portfolio management.

4.9 FUTURE TRENDS IN DATA WRANGLING FOR IIOT

4.9.1 Evolving technologies and techniques

The collection of raw data and organizing the dataset so that it is useful requires a lot of effort and time. Hence data wrangling is used to transform the raw data collected during the manufacturing process into more structured and easily interpretable data so that they can be used for further analysis. To transform the data, data wrangling requires six different steps:

a. *Data Discovery*: The first step is data discovery, which requires understanding the raw data and becoming familiar with it. It involves looking at the patterns and structure and configuring each dataset so that they can be understood and examined to find patterns and trends. This also requires removing errors and ambiguities so that it will be easier to derive information from the data.

b. *Data Structure*: The collected raw data can be of different formats and sizes and come from different sources. This data can be disorganized, so it needs to be restructured to be used as input in analytical models, such as machine learning models. Data structuring can involve removing text-heavy content such as dates, numbers, ID codes, HTML codes, etc. Other features include converting images to text, columns to rows, and transforming data into a more user-friendly spreadsheet format. This helps in the extraction of useful patterns and insights from the data.

c. *Data Cleaning*: Collected raw data can have many missing or repeated data. To remove these errors and redundancies from the dataset, this process involves determining what to keep and what to remove. Some of the data cleaning process involves removing rows or columns, addressing outliers, converting string characters to numerical values, handling null values, and deleting bad data completely. This process also involves fixing structural errors, correcting typos, and validating the data.

d. *Data Enriching*: As the size of the data can greatly determine efficiency and improve the accuracy of the analysis, we can combine data collected from different sources. This involves data augmenting, creating more subsets and variables, and adding more features from the existing dataset.

e. *Data Validating*: Data validation involves the use of automated processes to check the quality, consistency, accuracy, security, and authenticity of the data. This is achieved by checking dataset fields and attributes until the data is ready for analysis.

f. *Data Publishing*: The last step is now to publish the dataset for analysis. This data can now be available for the organization and company for analysis or can also be further processed to create a larger and more complex data structure as Data Warehouses.

4.9.2 Anticipated challenges and solutions

Data wrangling is used to transform raw data collected from sensors, devices, and other systems in the manufacturing equipment to a clean, usable form. The accuracy of the extracted information from the data highly depends on the data-wrangling process. However, in IIoT, data wrangling faces several challenges, such as the following:

a. *Scalability and Complexity*: With large sensors and devices used in IIoT, the data collected are massive. Hence, data wrangling needs to handle this big data in large volumes and quantities. It requires scalability and efficient tools, procedures, or advanced techniques to maintain this data for data ingestion, storage, and processing. Maintaining efficiency and speed while processing these huge amounts of data can be quite challenging. Additionally, different data coming from different sources needs to be handled for the data to be more efficient and effective. Structured, semi-structured, or unstructured data from these different IIoT sources, which may be in different forms such as text, images, videos, or social media posts, need to be integrated and harmonized. This adds complexity as it requires complex transformations, mappings, or conversions.

b. *Security and Privacy*: The data collected from the sensors or equipment in industrial sectors is highly sensitive. They need to be protected under various data privacy laws such as General Data Protection Regulation (GDPR) and the California Consumer Privacy Act (CCPA). This information may include product data, personal information, financial records, the performance of the system, and data collected through customer behavior, which are the intellectual property of the company and need to be protected from unauthorized users and access. Leakage or tampering of these data can cause a huge loss of property and reputation to the company. So, balancing the requirement of data wrangling while maintaining data security is one of the greater challenges faced during data wrangling.

To overcome these challenges, different data-wrangling tools, techniques, and best practices can be applied that are more robust and intelligent. For example, different encryption, authentication, or blockchain technology can be used for data security. State-of-the-art methods such as cloud computing, edge computing, or distributed systems can be used to ensure scalability and efficiency. These can be implemented through different data governance and data ethics frameworks for ensuring data privacy and security. Also, to enhance data quality, machine learning, data mining, or anomaly detection can be applied to improve data quality and generate more precise and actionable insights.

4.9.3 The role of artificial intelligence in data wrangling

Artificial intelligence (AI) plays a vital role in data wrangling for IIoT as it helps overcome the challenges of data security, privacy, volume, and complexity in IIoT data wrangling. AI can also enhance data quality and analysis, which can be beneficial for profits and production efficiency using IIoT data. Additionally, data wrangling provides clean, organized, and structured data that can be fed into AI models, such as machine learning for data analysis [16]. With technologies such as 5G, WiFi 6, edge cloud, digital twins, and smart factories, the huge data collected are analyzed to optimize processes, improving quality, efficiency, productivity, and turnover. AI-based IIoT platforms have helped achieve this by allowing companies to improve quality and reduce waste by fitting data into AI models. This aids decision-making and provides intelligent insights for predictive algorithms. AI, along with data wrangling, has shown applications in various fields such as smart manufacturing, prognostics and health management, power and energy, automotive and transportation, logistics, healthcare, communications and networks, smart cities, and automation [17].

AI can be used in data wrangling for transforming data formats, most likely from PDFs, audio files, and images. For example, AI can be used to extract data from PDF documents that provide information about products or services and use it for analysis. This makes the process faster than reading each line from the documents. Additionally, with data coming from multiple sensors in the manufacturing process, AI can determine from which sensors the faulty data is originating so that it can be repaired or sent for maintenance. Furthermore, for controlling manufacturing quality, computer vision can be used, such as in packaging film, to detect faults. AI can convert and analyze images of packages to identify any defects and make this information available to the relevant personnel in real time. These are some of the applications and advantages of using data wrangling along with AI.

4.10 CONCLUSION

With the number of sensors, devices, systems, and machines in industrial sectors adding each year with the development of Industrial IoT, data wrangling has become one of the important sectors that help to prepare this data for analysis. With these different sources sending enormous amounts of data, data wrangling helps to organize these unstructured data into clean, structured, and usable forms. This enormous data, after collection, needs to be transformed into a more suitable format involving data cleaning, data remediation, or data munging processes. These data are then analyzed and used for real-time decision-making and predictive analytics. They can also be used with AI or other key features of AI, including edge computing and machine learning. The findings can help improve operational efficiency, seamless connectivity, reduce costs [18], and improve product quality.

Although data wrangling is one of the crucial steps in the analytical process, it has faced some common challenges in handling missing data, ensuring data quality, maintaining data integrity and security, and seamless data integration. Also, with the enormous amounts of data in big data, it requires highly skilled manpower, processes, software, tools, and resources to maintain the data and extract insights that help optimize the manufacturing process. The development of AI, edge computing, cloud-based computing, and machine learning has made data wrangling one of the important factors, as the insights coming out are only as good as the data that informs them.

REFERENCES

1. S. Ullah et al., "A new intrusion detection system for the Internet of Things via deep convolutional neural network and feature engineering," *Sensors*, vol. 22, p. 3607, 2022, https://doi.org/10.3390/s22103607.
2. C. Bashya, M. N. Halgamuge, and A. Mohammad, "Review on analysis of the application areas and algorithms used in data wrangling in big data," *Cognitive Computing for Big Data Systems Over IoT. Lecture Notes on Data Engineering and Communications Technologies*, vol. 14, pp. 337–353, 2018.
3. M. Niranjanamurthy, K. Sheoran, G. Dhand, and P. Kaur, *Data Wrangling: Concepts, Applications and Tools*. John Wiley & Sons, New York, 2023.
4. Y. Sasaki, "A survey on IoT big data analytic systems: Current and future," *IEEE Internet Things Journal*, vol. 9, no. 2, pp. 1024–1036, 2021.
5. D. Mourtzis, E. Vlachou, and N. Milas, "Industrial big data as a result of IoT adoption in manufacturing," *Procedia CIRP*, vol. 55, pp. 290–295, 2016, https://doi.org/10.1016/j.procir.2016.07.038.
6. V. P. Gupta, "Smart sensors and industrial IoT (IIoT): A driver of the growth of Industry 4.0," *Smart Sensors for Industrial Internet of Things, Internet Things Challenges, Solutions, Applications*, vol. 2021, pp. 37–49, 2021.
7. R. Kodali, G. Swamy, and L. Boppana, "An implementation of IoT for healthcare," *2015 IEEE Recent Advances in Intelligent Computational Systems (RAICS)*, pp. 411–416, 2015, https://doi.org/10.1109/RAICS.2015.7488451.
8. S. Ahmad et al., "An Integration of IoT, IoC, and IoE towards building a green society," *Science Program*, vol. 2022, pp. 2673753, 2022, https://doi.org/10.1155/2022/2673753.
9. J. Astill, R. A. Dara, E. D. G. Fraser, B. Roberts, and S. Sharif, "Smart poultry management: Smart sensors, big data, and the Internet of Things," *Computers and Electronics in Agriculture*, vol. 170, pp. 105291, 2020.
10. J. M. Balajee and others, "Data wrangling and data leakage in machine learning for healthcare," *International Journal of Emerging Technologies and Innovative Research*, vol. 5, no. 8, pp. 553–559, 2018.
11. S. MacAvaney, A. Yates, S. Feldman, D. Downey, A. Cohan, and N. Goharian, "Simplified data wrangling with ir_datasets," *Proceedings of the 44th International ACM SIGIR Conference on Research and Development in Information Retrieval, ,* Canada, pp. 2429–2436, 2021.
12. L. Mazilu, N. Konstantinou, N. W. Paton, and A. A. A. Fernandes, "Data wrangling for fair classification," *EDBT/ICDT Workshops*, Australia, 2021.
13. S. R. Shakya and S. Jha, "Challenges in Industrial Internet of Things (IIoT)," in *Industrial Internet of Things*, CRC Press, Boca Raton, FL, 2022, pp. 19–39.
14. A. Arya and S. Jha, "An analytical review on data privacy and anonymity in 'Internet of Things (IoT) enabled services,'" *2021 3rd International Conference on Advances in Computing, Communication Control and Networking (ICAC3N)*, pp. 1432–1438, 2021. https://doi.org/10.1109/ICAC3N53548.2021.9725770.

15. J. Kazil and K. Jarmul, *Data Wrangling with Python: Tips and Tools to Make Your Life Easier*. O'Reilly Media, Inc., Sebastopol, CA, 2016.
16. J. Kepner, V. Gadepally, H. Jananthan, L. Milechin, and S. Samsi, "{AI} Data Wrangling with Associative Arrays," *CoRR*, vol. abs/2001.06731, 2020, [Online]. Available: https://arxiv.org/abs/2001.06731.
17. M. Elansary, Data wrangling & preparation automation. Why should you lose 80% of your time in one task?, 2021. https://doi.org/10.13140/RG.2.2.13981.03047.
18. A. T. Bogatu, *Cost-Effective Data Wrangling in Data Lakes*. The University of Manchester (United Kingdom), Manchester, 2020.

Chapter 5

Evaluating quality parameters for smart technologies in healthcare through social media platforms

Kamini, Tejinderpal Singh Brar, Rachit Garg, Rabie A. Ramadan, and Deepak Prashar

5.1 INTRODUCTION

Nowadays, due to the wide range of uses and popularity of social media platforms and applications, the demand for social media and other modes of communication has increased [1–5]. A wide range of communication modes are available personally and professionally, including storytelling, sharing on video sites, virtual chat rooms, and online web forums for discussing any topic. In this era, social media is emerging as an effective tool wisely used by people. However, the changing nature of this medium can impose issues for posting opportunities. It allows specific time for thought processes and what type of posted material on the internet is discoverable by law. The use of social media is embraced for personal use, including posting content on Facebook pages, Twitter, and blogs on various web links [6–10]. It has become crucial not to compromise privacy or disclose personal details, especially those of patients. In particularly challenging or emotionally charged situations, nurses may use e-portals and media to share personal experiences.

Undoubtedly, social media plays an indispensable role for organizations, given its rapid growth in popularity and industry channels. Nevertheless, a clear approach to the mode of external communication as a social media interaction tool is generally lacking. Both internal and external communication can be enhanced through the intensive use of social media. However, much of the research findings focus on external modes of social media. Currently, healthcare organization has been using social media for communication with external environments. The expanding market and the increasing demand for healthcare departments have made it imperative to focus on this area. It becomes imperative for patients to participate in healthcare discussions to make informed decisions [11–15]. In order to attain the requirements of customers, many social network platforms are available for sharing and accessing health-related knowledge. This study also facilitates healthcare professionals to delve deeper into understanding the role of online social platforms for their best optimization.

DOI: 10.1201/9781003530572-5

5.1.1 Literature review

Objective	Scope	Data
To examine how big data impact healthcare services	Big data and healthcare	Christodoulakis, A., Karanikas, H., Billiris, A., Thireos, E. & Pelekis, N., 2016, "Big data" in healthcare assessment of the performance of Greek NHS hospitals using key performance and clinical workload indicators', *Archives of Hellenic Medicine* 33(4), 489–497. Kayyali, B., Knott, D. & Van Kuiken, S., 2013, '*The big-data revolution in US health care: Accelerating value and innovation*', Mc Kinsey & Company 1–13. Schouten, P., 2013, 'Big data in health care: Solving provider revenue leakage with advanced analytics', *Healthcare Financial Management* 67(2), 40–43. Feldman, B., Martin, E.M. & Skotnes, T., 2012, 'Big data in healthcare hype and hope', October 2012. Dr. Bonnie, 360, October 2012, viewed 10 November 2016, from https://www. ghdonline.org/uploads/big-data-in-healthcare_ B_Kaplan_2012.pdf
To examine the role of integration in the security of healthcare data	Healthcare data security and privacy	Al Ameen, M., Liu, J. & Kwak, K., 2012, 'Security and privacy issues in wireless sensor networks for healthcare applications', *Journal of Medical Systems* 36(1), 93–101. Hsu, C.L., Lee, M.R. & Su, C.H., 2013, 'The role of privacy protection in healthcare information systems adoption', *Journal of Medical Systems* 37(5), 9966. Mišić, J. & Mišić, V., 2008, 'Enforcing patient privacy in healthcare WSNs through key distribution algorithms', *Security and Communication Networks* 1(5), 417–429.
To understand the role and effect of social media in the delivery of healthcare services	Social media and healthcare	Sarasohn-Kahn, J., 2008, 'The wisdom of patients: Health care meets online social media', viewed 11 November 2016, from https://pdfs.semanticscholar.org/5301/8d4afd6a 01edd3a22dc3e062b3a29fb4e222.pdf Househ, M., Borycki, E. & Kushniruk, A., 2014, 'Empowering patients through social media: The benefits and challenges', *Health Informatics Journal* 20(1), 50–58. Vance, K., Howe, W. & Dellavalle, R.P., 2009, 'Social internet sites as a source of public health information', *Dermatologic Clinics* 27(2), 133–136. De Choudhury, M., Gamon, M., Counts, S. & Horvitz, E., 2013, 'Predicting depression via social media', in International Conference on Web and Social Media, Cambridge, MA, July 8–11, pp. 128–137.

5.2 TOOLS FOR DISSEMINATING INFORMATION TO PATIENTS

Twitter: It is the most vibrant platform in the health industry to connect and update people around the globe. So, it becomes important to get in touch with the right choice people on Twitter who can help in the future and change their careers. However, as a medical professional, Twitter also helps personally and professionally for business purposes. There is integration of networks with user-generated pages to help prompt any question. Twitter's social source is used for the objective of advanced marketing tools and methodologies and might not be suitable for everyone [16–18].

Social interaction using Twitter, with online posts known as "tweets", is limited to 140 characters. The Twitter service has 500 million active users who write and read these tweets. This free service has active users generating over 340 million tweets and 1.6 billion search queries per day. Twitter has also become one of the top 10 most visited sites on the internet, generating traffic to websites and aiding in sales. It is recommended for more passive research and recreation [19–21]. Globally, there are 319 million active users, with 100 million accessing it daily via smartphones. Individual tweets can reach up to 500 million users, and during live broadcasts, three out of five television shows interact with their viewers. Original tweets engage a higher number of users as compared to copied tweets.

Facebook: In foreign countries like the United States, most people search for healthcare services either on Facebook or mobile apps for checkups and preventive measures, including cholesterol tests. Recommendations are based on age and gender details provided by users. Notifications regarding flu shots are also provided timely.

An online tool is also available to remind people about their schedules, share ongoing test reports with loved ones, and increase awareness among the younger generation about preventive measures. The other objective is to find places to receive care at affordable rates.

Blogs: Started by Modern Medicine Network, this system helps physicians practice blogs to share knowledge and experiences about emerging topics related to healthcare. With respect to each process, each article in the blog section includes topics that are extremely relevant to healthcare professional's work and experience. Additionally, blogs encourage the audience to showcase their expertise in the healthcare industry. Instilling trust and building strong relationships also results in better business intelligence.

YouTube Patient Testimonials: There is a communication tool with the use of video media content and sharing forms, which may not be substantial for marketing in healthcare branches. However, other media-sharing platforms such as YouTube Spectrum are valuable source for marketing purposes. On the contrary, TV and audio-visual modes of commercial marketing techniques find proactive users searching for content and bring the opportunity to reach interested users and large audiences from distinct groups (Figure 5.1).

Figure 5.1 Real-time updates through social media.

LinkedIn: When a user wants to connect with other healthcare professionals in the same area of specialization or different groups, LinkedIn groups help achieve this target. The other options available in the search include student and doctor relationships, medical device connectivity, practice networks for healthcare physicians, nurse, and staff availability for group discussions, and professionals in pharmaceutics.

5.3 HEALTHCARE PROMOTION STRATEGY

1. *Clients Are Searching for Doctors Online*

   The statistics don't lie: 90% of consumers are searching online for businesses to do their research before deciding to work with them. While that's true of all sorts of companies, remember that a doctor's office is unique. Think about the kinds of people who are likely searching for them:
   - Families who have relocated to the area and are looking for a family practice.
   - People who are trying to find a new doctor for one reason or another.
   - Customers thinking about moving to the area and falling into the above 90%.

In recent times, Instagram has been occupied with publicity for online businesses for various products and services compared to content on social network websites, helping achieve the current marketing trends and goals. From the survey results, it is clear that 71% of businesses use Instagram to market their services, and the site boasts more than 25 million active businesses dynamically. In another finding, it is clear that 80% of Instagram accounts include at least one business, and 30% of users have buying behavior from the discovery data on Instagram. It is also stated that Instagram's marketing policy is effective for convincing private practices.

5.3.1 Design an Instagram profile that informs and reflects your business

To create a business profile, it is mandatory to convert the current state of the Instagram profile to a business level. This allows for extracting features to measure the success rate of Instagram marketing efforts and providing direct connectivity to the office by inheriting the profile of the patients. In addition to this, when choosing an Instagram username (@youruser-name), try to avoid directly picking your name. Furthermore, for business enhancement, add a high-quality profile image. The recommendation is a professional headshot with a company logo, depending on the branding department. Finally, don't forget to add information to your biography. Basic information such as name, location, and place are required to complete the profile. Catering to patients through a link provided by the app is also an efficient way of service.

5.3.2 Top healthcare app for the year 2019/2020

- Concord Cloud Fax
- HIPAA Security Risk Analysis Software
- InDxLogic
- Opargo
- Optum Patient Insights
- Charlie by Healthfinch
- Relaymed
- ChartScout
- Point of Care AC
- Vital Signs Device Integration by Hill Rom (Figure 5.2)

As per recent research conducted by Google, 84% of patients use both online and offline social media sources for hospital research. Additionally, search drives nearly 3× as many visitors to hospital sites compared to non-search visitors. Patients primarily search for the latest symptoms and condition terms leading up to the moment of conversion. Furthermore, 44% use their mobile phones for scheduling appointments.

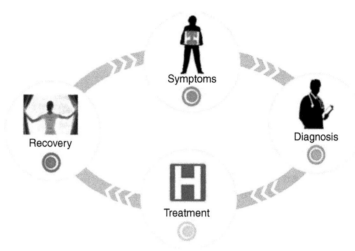

Figure 5.2 Patient journey state diagram.

5.4 TYPES OF CONTENT TO PUBLISH ON SOCIAL MEDIA

A few content ideas to consider include:

- *Research Summaries*: If you've recently published new research, consider sharing snippets of that information on social media.
- *Blog Posts*: Publish blog posts on your website and then share links to those blog posts on your social media profiles. Blog posts can answer questions that patients frequently ask, discuss your treatments and services, or talk about timely topics.
- *Photos and Videos*: Showcase your staff and office space in photos and videos. Also, consider showing off your patients, with their written permission.
- *Media Articles*: If you are quoted in a media article, share that too. Even though the article is not original to you, your quote is. Plus, you can add more context to the article in the post itself.
- *Educative Content*: The ongoing trends of the self-diagnosis system make it possible to educate the public through Google services. Providing social care services to customers through practice also helps educate potential patients. The objective is to provide timely information about treatment options and practice purposes.
- *Meaningful Content*: Contents related to progressive diseases such as cancer can be impactful. Family members mostly use inspirational and motivational content in such cases.

The audio-visual content appears on Instagram to represent images and showcase creativity in current market practices. On Instagram, try using stories to go behind the scenes, share images of your attendance at networking events, or feature new products and services.

Some tips for the type of content on Instagram:

- *Get Personal*: People love stories, so tell yours. Do more than share a link or video; explain what the content means to you and why you're posting it.
- *Use a Compelling Quote*: Let's say you publish a blog post in which a woman shares her experience with a recent surgical procedure and how it has changed her life. Pull a gripping quote from the article and add depth to your post.
- *Ask an Engaging Question*: "Do you agree with...?" "Would you rather...?" "Have you ever...?" These questions are a strong start to engaging posts.

Remember that Instagram is a photo-sharing app. Therefore, your caption is primarily there to add context to the photo you're sharing and shouldn't be too long or take the focus away from the image you've shared.

5.4.1 Use of hashtags

Hashtags are used on all social media websites. They are descriptive in nature and can be clickable keywords. For example, clicking on the hashtag #beach will likely show you oceanic views from across the world.

Hashtags are most popular on Twitter and Instagram and can help people find your posts. Twitter can be used to engage with your peers in the medical community; many common hashtags like #PatientExperience and #physicians are often used by thought leaders in the health space.

The users can create their own hashtags for their own logos or brands by simply placing the pound symbol (#) in front of them. Tags created by healthcare professionals, such as #dentisttips, can become something patients follow or look forward to seeing.

During a big news event, look for the hashtag(s) associated with it. If there's a new breakthrough in cancer research or a viral outbreak, you can attract a new audience seeking your expert commentary.

On Twitter, results are best seen if you limit your tweets including one hashtag. 16 On Instagram, it can be tempting to use more than 20 hashtags to increase the visibility of your post, but limit your hashtags to about 5–10.

Posts with nine hashtags tend to see the most engagement on Instagram.

5.4.2 Scheduling content in advance

Scheduling posts in advance can help streamline your approach to healthcare social media. Some social media management platforms allow you to schedule posts to multiple social sites simultaneously and keep track of any messages you may be receiving. Some of the most popular platforms include HootSuite, Later, and Buffer.

5.4.3 Length of posts

The only social media channel with strict character limits is Twitter. Historically known for its 140-character limit, Twitter doubled its limit to 280 characters in 2018, allowing your tweets to be a couple of sentences long. This means you may have to worry less about your posts getting cut off by Facebook and Instagram.

5.5 PATIENT ENGAGEMENT AND SATISFACTION

Other social media networks such as Facebook groups can enhance patient engagement and satisfaction according to a published research study.

As per recent studies and reports, incorporating social media into practical concepts using Facebook provides substantial benefits to numerous patients, with almost 90% of survey respondents giving positive feedback based on the results retrieved. Additionally, sustainable results were attained, with varying numbers between 10% and 15% monthly, and reports remained constant throughout the entire study.

With the use of virtual community interaction forum systems, many research findings show the Facebook group for liver transplant patients. Participation was from different areas such as caregivers, family members, and health service providers. Out of 350 users, 50% of respondents were from liver transplant services areas, 36% were from caregivers and friends, and 14% were healthcare providers.

The activity of posts and reactions were recorded for active users in the group, and the contribution ratio was 64 and 61, respectively. The results showed that the most common type of post was based on supportive and inspirational content.

At the end of the study, participants were asked to fill out a form, and 63 out of 350 respondents replied. The demographic group selected 75% of the females out of 90%, that resulted into the major consequence.

Patients were most likely to participate in sharing supportive posts and comments. Many patients shared advice and experiences on social networking groups. The main reason for participating in the group was to receive and provide social support from various channels.

"These patients share common stressors related to experiencing physical and psychosocial symptoms associated with end-stage liver disease, dealing with the uncertainty of waiting for an available organ, and managing strict medication regimens," the authors in the study wrote. "Our findings support this notion, because 44 percent of survey respondents reported they were unlikely to participate in a similar group for primary care or other subspecialties because they felt the online support group was unique to transplantation."

The study materials shared by daycare staff were also more likely to emphasize empathy rather than simplicity. According to a recent survey conducted by the healthcare department, 82% of the content was exploratory and descriptive.

For conducting the study, Facebook users were deployed, with almost 70% giving responses through the Facebook page survey form daily, and 25% of patients reported visiting the page multiple times a day.

However, there is a limitation in the study as the number of survey takers was limited. In future studies, more patients and healthcare professionals will be involved to evaluate the impact of social media initiatives for post-transplant care patients.

The current work contributes to the enhancement of social media impact, particularly for liver transplant and surgical patients through personal engagement and satisfaction. With the growing expansion of social media, many webinars and live programs related to the healthcare surgical system begin by embracing the delivery of lectures and seminars on patient care, greatly impacting medical departments and survivors.

5.5.1 Social media reviews make all the difference

The uploaded reviews on social media are prominently visible on sites like Yelp and Vitals. Users prefer to visit and interact with the Facebook page daily rather than Yelp. The social media content will boost and strengthen patient engagement, providing opportunities for conversations with guardians. It also increases exposure and strengthens ranking results. There is also promotion of team building among staff members.

Social media has a direct impact on your reputation. The Google search algorithm helps boost search rankings and improve social posting, commenting, and responding.

5.5.2 Monitoring social media success

When you're investing time, energy, and money into a social media marketing strategy, it's important to measure the results of your efforts. Facebook, Instagram, and Twitter all give users insight into how their posts are performing. If you're using a social media management system, such as HootSuite, you can view analytics for all your accounts on one platform.

We recommend tracking the following metrics:

- *New Followers*: Take note of your followers. Is your audience growing? If so, keep track of how many new followers you are receiving every month. This will give you an idea of the rate your accounts are growing. That way, you'll know when you have a particularly great month or a slow month.

- *Impressions or Post Reach*: Impressions track how many people have seen your posts. This is an important metric to note because, to be effective in healthcare marketing, you need more eyeballs on your content and more people to be aware of who you are. Most social media marketers (70%) list increased brand awareness as their top goal for social media, and tracking impressions is a good way to measure awareness.
- *Engagement*: An engagement is an action someone takes on your content. This includes liking, commenting, and sharing, but it could also include swiping through a photo slideshow or watching a video you've posted. An engagement shows that someone did not just scroll past your post but is an indication that people are interested in the content you're posting. Monitoring your engagements can let you know whether your content is resonating with your audience.

5.6 BARRIERS TO USING SOCIAL MEDIA FOR HEALTHCARE

- *Damage to Professional Image*: Healthcare professionals must be mindful of the information and photographs they circulate in the media, adhering to time limits. The ethics regarding sharing posts on social media must be followed positively, respecting the code of conduct rules and national laws.
- *Licensing Issues*: Every hospital and clinical center is required not to share any private or personal authentication with a third party about a patient's illness without the patient's permission. There should be clear content or information stated in the post when permission is assigned to any doctor. A signed document should state that uploaded information or opinions represent the individual's own content, not other personal data. In case of any legal issues, according to law regulations, content without permission should not be disclosed on social media, even depending on the purpose for which the information is required.
- *Poor Quality of Information*: There is a lack of reliability in the information found on social media. Moreover, several authors have found that social media sites often contain limited information. Additionally, medical information was found to be either incomplete or informal. Similar issues exist with traditional online media; however, it is essential to note that the interactive nature of social media aggravates these issues manifold as any online user can upload unauthorized content to any site.
- *Unprofessional Conduct*: Recently, it has become extremely urgent to maintain professionalism along with personal boundaries in online social environments with patients. Open discussions with patients regarding health issues on social networking platforms such as Facebook, Twitter, or WhatsApp are strictly prohibited, although healthy discussions on any topic are always encouraged. While

keeping in regular touch with doctors shows patient interest in discussions, distance needs to be maintained for personal opinions.

- *Return on Investment Barriers*: There are numerous ways of marketing, but in an online environment, social media is an effective tool, and its success starts with identifying services and developing an appropriate strategy. An advantage of using social media tools is that it is mostly free, though it can take significant staff time to engage. There are associated costs such as the time spent by staff away from other projects. However, well-planned social media projects can attract new and loyal patients. Previous research studies show that connecting with a hospital through social media is an effective way for patients to decide to seek treatment at that hospital.
- *Time Barrier*: In today's fast-paced era, time is precious for healthcare professionals and other healthcare providers. Keeping up with every Twitter or Facebook update is time-consuming and almost impossible. It is better to organize incoming information and create lists that allow you to focus on certain themes rather than read every update. For example, organize by specialty or level of importance.
- Do not try to do everything; address those societal needs that you think are most important, or that motivate you. Using social media during your free time undoubtedly brings added responsibility, so focusing on issues that are relevant to you makes it easier to sustain the effort.
- *Legal and Regulatory Barriers*: It is important for healthcare professionals and providers not to discuss details of patients' health conditions or illnesses or any personal information on online platforms without the patients' consent.

5.6.1 Recommendations/Future considerations

- To understand the intention behind user keyword phrasing, the concept of natural language processing can be used.
- For answering conventional queries, Google is getting better and better at understanding longer sentences.
- Due to advancements in technology, search engines like Google can now sense the importance of prepositional phrases that contain words like "of" or "to" for meaning.

5.7 CONCLUSIONS

ICT-based social media platforms have incredible power for both intensifying personal relationships and offering precious and timely information to healthcare consumers. They provide a shareable opportunity to interact with colleagues from all corners of the world. To apply social and economic stability through electronic social media, it is extremely significant for healthcare organizations and authorities to be aware of the potential costs of disclosing patient-related queries and databases through electronic platforms. There have been personal and professional benefits of social media platforms without affecting privacy and policy matters.

REFERENCES

1. Albert, S. & Whetten, D.A. (1985). Organizational identity. In Cummings, L. L., & Staw, B. M. (Eds.), *Research in Organizational Behavior* (pp. 263–295). Greenwich, CT: JAI Press.
2. Kietzmann, J.H., Hermkens, K., McCarthy, I.P. & Silvestre, B.S. (2011). Social media? Get serious! Understanding the functional building blocks of social media. *Business Horizons*, 54, 241–251.
3. Hutton, J.G. (1999). The definition, dimensions, and domain of public relations. *Public Relations Review*, 25(2), 199–214
4. Iyyappan, M., Kumar, A., Ahmad, S., Jha, S., Alouffi, B. & Alharbi, A. (2023). A component selection framework of cohesion and coupling metrics. *Computer Systems Science and Engineering*, 44(1), 351–365.
5. Holtzhausen, D.R. & Roberts, G.F. (2009). An investigation into the role of image repair theory in strategic conflict management. *Journal of Public Relations Research*, 21(2), 165–186. https://doi.org/10.1080/10627260802557431
6. Fombrun, C.J. & van Riel, C.B.M. (1997). The reputational landscape. *Corporate Reputation Review*, 1(1), 5–13.
7. Arya, A. & Jha, S. (2021). "An analytical review on data privacy and anonymity in "Internet of Things (IoT) enabled services," *2021 3rd International Conference on Advances in Computing, Communication Control and Networking (ICAC3N),* Greater Noida, India, pp. 1432–1438, https://doi.org/10.1109/ICAC3N53548.2021.9725770.
8. Ahmad, S., Jha, S., Alam, A., Alharbi, M. & Nazeer, J. (2022). Analysis of intrusion detection approaches for network traffic anomalies with comparative analysis on Botnets (2008–2020). *Security and Communication Networks*, 2022, 9199703.
9. Singh, K. & Jha, S. (2021). "Lean manufacturing - An analytical approach towards Industry 4.0," *2021 2nd International Conference on Smart Electronics and Communication (ICOSEC), Trichy, India,* pp. 1690–1695, https://doi.org/10.1109/ICOSEC51865.2021.9591684.
10. Daft, R.L. & Lengel, R.H. (1984). Information richness: A new approach to managerial behavior and organizational design. In Cummings, L. L., & Staw, B. M. (Eds.), *Research in Organizational Behavior* (Vol. 6, 191–233). Homewood, IL: JAI Press.
11. Chou, W.-Y.S., Hunt, Y.M., Beckjord, E.B., Moser, R.P. & Hesse, B.W. (2009). Social media use in the United States: Implications for health communication. *Journal of Medical Internet Research*, 11(4), 1–12. https://doi.org/10.2196/jmir.1249
12. Hawn, C. (2009). Take two aspirin and tweet me in the morning: How twitter, facebook, and other social media are reshaping health care. *Health Affairs*, 28(2), 361–368. https://doi.org/10.1377/hlthaff.28.2.361

13. Ahmad, S., Jha, S., Alam, A., Yaseen, M. & Abdeljaber, H.A.M. (2022). A novel AI-based stock market prediction using machine learning algorithm. *Scientific Programming*, 2022(1), 1–11.

14. Picazo-Vela, S., Gutiérrez-Martínez, I. & Luna-Reyes, L.F. (2012). Understanding risks, benefits, and strategic alternatives of social media applications in the public sector. *Government Information Quarterly*, 29, 504–511.

15. Pinterest. (2013, January 10). Wikipedia, the free encyclopedia. Retrieved from https://en.wikipedia.org

16. Tang, X. & Yang, C.C. (2012). Ranking user influence in healthcare social media. *ACM Transactions on Intelligent Systems and Technology*, 3(4), 73:1–73:21. https://doi.org/10.1145/2337542.2337558

17. Tucker, L. & Melewar, T.C. (2005). Corporate reputation and crisis management: The threat and manageability of anti-corporatism. *Corporate Reputation Review*, 7(4), 377–387.

18. Bezner, S., Hodgman, E., Diesen, D., Clayton, J., Minkes, R., Langer, J., & Chen, L. (2014). Pediatric surgery on YouTube(tm): Is the truth out there? *Journal of Pediatric Surgery*, 49, 586–589.

19. Jha, S., Jha, N., Prashar, D., Ahmad, S., Alouffi, B. & Alharbi, A. (2022). Integrated IoT-based secure and efficient key management framework using hash graphs for autonomous vehicles to ensure road safety. *Sensors*, 22(7), 2529.

20. Moreno, M., Grant, A., Kacvinsky, L., Moreno, P. & Fleming, M. (2012). Older adolescents' views regarding participation in Facebook research. *Journal of Adolescent Health* 51, 439–444.

21. Knight, E., Werstine, R., Rasmussen-Pennington, D., Fitzsimmons, D., & Petrella, R. (2015). Physical therapy 2.0: Leveraging social media to engage patients in rehabilitation and health promotion. *Physical Therapy*, 95, 389–396.

Chapter 6

Security standards of IIoT

Mohammad Shuaib Khan and
Mohammad Mazhar Afzal

In the current circumstances, the Industrial Internet of Things (IIoT) security poses a significant challenge. It goes without saying that employees will use laptops, smartphones, and tablets known as "bring your own devices" or "BYOD" devices to communicate with one another and complete tasks within the organization. However, the provision of IIoT and network security standards presents a significant obstacle for IT professionals. Nobody is aware of the organization's connections, which pose a significant threat. If the IT staff is unable to account for risks like invisibly connected infiltrators, how will they safeguard sensitive information from hackers?

We can broadly classify the various threats to the organizations' security as

1. Administrative and operational risks
2. Technical risks
3. Physical risks

IIoT security allows the Chief Information Security Officer, Original Equipment Manufacturer, and administrators of the network to protect and manage the connected device endpoints on their network. This includes the following:

- *IIoT Endpoints*: The use of internet-connected devices for data analysis and predictive maintenance increases the number of endpoints that could be attacked.
- *Remote Connections*: Many IIoT devices are now being accessed from unreliable and possibly unsecure Wi-Fi networks as remote work becomes more common.
- *Legacy Devices*: Older devices frequently cannot be patched, introducing vulnerabilities known as "low-hanging fruit."

An IIoT network breach could have devastating effects. Ransomware attacks that shut down entire production networks or leaks of proprietary

DOI: 10.1201/9781003530572-6

data and information in critical infrastructure industries are examples of these consequences. The Internet of Things (IoT) has been welcomed with great enthusiasm by the manufacturing sector. Process efficiencies, increased cost-effectiveness, and product innovation are driven here by IIoT. This is the presentation of a new network, investigation stages, and shrewd sensors for assembling production line floors' modern settings.

A stringent set of operating standards and regulatory frameworks also govern the manufacturing sector. These are necessary to maintain quality, health and safety, the environment, data protection, and privacy, among other things, at acceptable levels. However, the particular norms and structure that will apply are still in flux since the IIoT is still a young, unproven, and constantly emerging approach, with particular difficulties associated with its mechanism and the organizations involved.

To guarantee minimum quality standards and appropriate interoperability in any setting, standards between various organizations and sectors are essential. Because it ensures both scalability and growth, this interoperability is especially important in the context of the IIoT.

Hence, the prime concern is the overall security of the organizations within the proven standards of IIoT.

IIoT is a key enabler for smart manufacturing and industry that correlates and incorporates automated control systems with striving systems, the internet, business processes, and analytics. It has greatly expanded the variety of technologies that can be utilized in industrial settings. This functional innovation union presents new security dangers and difficulties that modern clients should suitably manage. Amazon Web Administrations prescribes a diverse approach to securing the modern control framework/functional innovation, IIoT, and cloud environments to help organizations arrange their digital transformation safely.

Almost everything is interconnected, from automobiles, refrigerators, and industrial machinery to wearable devices and cell phones. IT professionals have known for a long time that global connectivity will only increase rapidly. Our jobs are made more convenient by connectivity. Employees will almost certainly communicate using BYOD devices like laptops, smartphones, and tablets. However, the provision of IIoT and network security standards presents a significant obstacle for IT professionals. Not realizing what is associated with your venture's network is hazardous. If you and your IT staff are unable to account for risks like invisibly connected infiltrators, how will they safeguard sensitive information from hackers?

Sometimes it takes a village to be vigilant. Currently, there is a rush to establish universal IIoT standards that will bind an ample area of topics, along with security. Implementing global security standards for the IoT will unquestionably help your IT department prevent dangerous devices from wreaking havoc in your business.

Universal IIoT standards have been proposed by several network security standards groups, IoT architecture, interoperability, privacy, and security are all under consideration by these groups, but none of them has emerged victorious. It's like the famous VHS-Betamax battle; may the best ideas prevail. Most nations have been chipping away at creating individual IIoT guidelines, yet thoughts on the most proficient method to do that contrast.

Your IT manager probably noticed some recent signs of progress: Two brand-new global recommendations for IoT were created in Singapore by members of the International Telecommunication Union Standardization Sector (ITU-T) Study Group. One of these recommendations identifies a familiar magnitude for security management and software upgrades. If these issues were covered by industry standards, managing IIoT applications and devices would be simpler and more effective for your IT administrator.

Strong network access control is the answer for businesses that want to manage IoT. Network access control (NAC) permits associations to control who gets access to the Local Area Network and what activities they are allowed to perform once connected.

Numerous technological advancements and opportunities have emerged since Industry 4.0's inception. IIoT, which connects people, machines, devices, and sensors, is one of the most important themes of the Fourth Industrial Revolution. Network administrators and chief information security officers (CISOs) face new challenges in protecting their networks' devices from malicious attackers [1].

Employees in today's distributed workforce frequently connect remotely to IoT devices via home and public Wi-Fi networks, increasing the attack surface and cyber risk. Network administrators and CISOs require security [2,3] technologies that combine remote access and security without the historical difficulties of complex deployments or excessive overhead.

Why have standards not been implemented sooner? Criteria are very important for IIoT's continuous improvement as well as its future stability and dependability.

6.1 COMPETITION

In a highly dynamic and complex industry, various developers, suppliers, and vendors are fighting for elevation and intellectual property rights. As a result, establishing common standards is not always a top priority.

6.2 INTEROPERABILITY

One of the most important reasons for IIoT standards is to ensure interoperability, as previously stated. However, it also makes adoption difficult. IoT combines two distinct technologies, operational technology (OT) and information technology (IT), which means that there are more standards and more interoperability than in nonindustrial settings.

6.3 SECURITY

By implementing IoT, businesses significantly expand their attack surface. As a result, to safeguard these organizations from malicious cyber-attacks, robust processes for device verification and data protection must be formalized in yet another set of standards. Almost half of OT experts are concerned about the ability to correlate IIoT devices, according to the 2018 SANS Industrial IoT Survey. Until now, globally recognized and certified security architecture for the IIoT is lacking. It is difficult to navigate all this complexity, and all sector organizations [5,6] must be prepared for change until an international agreement is reached on a shared IIoT standards framework. There are certain steps that you can follow to ensure that you are applying dependable IIoT solutions and that you are strategically established for the future landscape.

Familiarize yourself with the landscape. The emerging standard landscape contains numerous methods, organizations, and people working to be attentive, as outlined in this document. Although it won't be possible for any organization, especially a smaller one, to keep an active eye on all the evolving objects [4], going to touch some similar organizations and fully take an active role in developing the standards scene should help you stand out in the future.

Most of the time, IIoT project managers are in charge of connecting the information systems engineers, the IIoT security ecosystem, and constantly changing classical prospects. There are various rotating components here, so careful management and a holistic but strategic approach are required. Interoperability ought to improve as a consensus on standards emerges over time [6]. In the meantime, be aware of the pressures internal and external project managers face and the sometimes-adverse preferences they must manage.

Security and interoperability should always be at the center of your IIoT strategy because they will be the guiding principles for any new standards. When a standards framework is agreed upon, you are much more likely to be in a good position if you incorporate them into your IIoT ecosystem from the beginning. The IIoT scene is in steady development, and change is currently the only certainty [7,8]. Notwithstanding, with a more profound comprehension of the difficulties and tensions confronting the improvement of IIoT norms, your organization can position itself in the best possible way for a more guided future.

Categorization of common security risks can be done as follows:

1. Administrative and operational risks
2. Technical risks
3. Physical risks

6.3.1 Administrative and operational risks

This sort of chance implies human-created gambles, if coincidental or purposeful, for example, mistaken utilization of IIoT gadgets or an aggressor attempting to break into the organization. The entire organization is at risk because there isn't an overall IIoT security risk administration plan in place. The program should incorporate reported approaches, methodology, and customary preparation to distinguish, oversee, and destroy network protection chances where conceivable.

Mishaps and blunders made by people can leave holes in security, such as through misuse or erroneous installation of IIoT devices with legacy frameworks being used. IIoT devices are also targeted by malicious insiders and outsiders of the organization. While organizations continue to make use of their IIoT technology, manufacturers of IIoT devices and systems may cease production, go out of business [8], or cease providing support. This may leave IIoT devices with vulnerabilities that can be exploited later and could be used for critical purposes.

6.3.2 Technical risks

The attack surface is expanded by adding more devices. Users may not be aware of the vulnerabilities in their devices, but cybercriminals can find them. Because IIoT formation establishes new relations amid IT and OT, deployment complications and security risks increase, especially for IIoT hardware, software, and firmware that are not standardized. Devices, controllers, and supporting systems suffer from inconsistent security due to the absence of internationally accepted technical standards for IIoT security and interoperability. Numerous IIoT devices use delicate or no verification, operate on weakly coded software that is susceptible to accomplishment, and use weak or no cryptography.

6.3.3 Physical risks

The operation of IIoT systems can be adversely altered or interrupted by natural disasters, events, or attacks. An attacker may target IoT devices that control resources like oil pipelines and make direct modifications to them that have a negative impact on the physical world. Every IIoT installation should have a set of cybersecurity objectives. Utilizing secure IIoT devices and systems is the first step. Devices must be configured by IoT teams to prevent them from being used in attacks like distributed denial of service, data exfiltration, or device settings modification.

Businesses must also establish data security controls. All data collected, processed, stored, or transmitted to or from IIoT technology must be secured to ensure its confidentiality, integrity, and availability. The inherent risks posed by IoT include inconsistent deployments and unreliable data [9].

Every organization applying IIoT devices must contain security potential in the following domains.

Security measures for operations and administration. These people-based actions are necessary to ensure effective IIoT deployment. The steps essential to control the selection, advancement, utilization, and upkeep of devices are included in the controls. Data and device functions from the IIoT must be safeguarded by security measures, as must workforce management via IIoT devices [10]. Documentation of IIoT policies and procedures, security control settings, secure-use training for devices, and IIoT risk assessments are among these measures [11].

6.3.4 Safety in technology

The technology-based parts are needed to protect the technological parts of IIoT system, such as cloud services or supply chains, and ensure effective IIoT security [12]. This includes controls like encryption, secure boot, IIoT device authorization with multifactor authentication, and device identification methods.

6.3.5 Physical protection

The IIoT environment's facilities, servers, controllers, and devices themselves are safeguarded by physical measures and tools [13]. Organizations are required to establish IIoT device and facility proximity access controls that restrict modification of IIoT device and system controls and allow only authorized access to proprietary data.

6.4 MAJOR IIOT SECURITY CONCERNS

The persons solely responsible for controlling and managing the IT should have to absorb IIoT devices, controllers, and supporting equipment into their overall area of network execution. Various methodologies exist, with major focuses on testing and activities related to risk management, vulnerability assessment, testing, data restoration as needed, and most important is the security training for all those who are utilizing IIoT devices and other subcomponents of the security parameters needed in such type of security program. Organizations can also adapt and utilize the seven sections of Cyber Security Framework (CSF) to construct and frame their security methodology and broaden security commands to add IIoT components, operations, and clouds.

The administrators of IT have to ensure that the most elementary approaches are already in place. Industries and organizations worldwide are effectively utilizing the cybersecurity framework, which helps improve the typical infrastructure cybersecurity of the organization.

Other trustworthy frameworks, such as ISO, ISA99, and IACSS, are also available for organizations.

For organizing and accommodating their security methodologies, industries can customize and even utilize the most popular seven CSF categories.

1. Asset management
2. Business environment
3. Governance
4. Risk assessment
5. Risk management strategy
6. Supply chain risk management
7. Information protection processes and procedures

As IIoT revolutionizes former systems related to the activities in our daily lives and businesses, its extension into physical spaces merges known digital threats within the cybersecurity policy realm. This affects the real world, with consequential susceptibility related to public safety, physical damage, and calamitous systemic threats to general popular frameworks. Bad actors are always seeking to exploit the digitally networked surroundings, breaking into houses with consumer devices, and penetrating the growing interlinked floor of the manufacturing and core construction sectors. The possible brunt of a threat to the analytical framework would be far from embracing. Health, safety, public welfare, and majorly national security are the critical considerations. The crucial challenge for objects, systems, and services relying on the IIoT is not the technology itself.

During the recruitment of network members from the entire industry, global organizations, civil societies, and academicians to examine the administrative structure, gaps in IIoT security, perks, and fines coordination that would enhance IIoT security practices, success, can be maximized. The effect of this task will be conducted by the consecutive considerations enclosed during scoping deliberations in October 2016.

The network should ensure broad stakeholder representation. Recognized academicians and technical industries indulge in discussions about the security of the IIoT. Following the recent high-profile breaches, public policy experts have participated in discussions, becoming aware of the severity and complexity of the problem. Prior to these incidents, IIoT users were not well informed. It is the responsibility of the organizations that they should not only to implement but also to circulate the strategies for IIoT security policy for the common good. IIoT security issues contribute to informed decision-making by all stakeholders involved.

The ideas about IIoT security concepts should be enhanced through network expansion. User knowledge of IIoT security issues tends to have less experience in handling security gaps. This applies across security communities and vertical markets, from small-scale units to large technologists. The ultimate goal is to simplify IIoT device security, ensuring interconnected components and sophisticatedly connected sensors in information technology can update their login credentials. Additionally, because of insufficient knowledge, IIoT users generally face problems, attributing minimal security to potential breaches. If the administrators fail to involve stakeholders, third-party vendors may face significant market failures.

Security parameters should be elaborated to firms and users through networks. Cybersecurity experts can be employed either by users or vendors. IIoT fails to explain its cybersecurity vulnerabilities or deploy long-term business strategies. Systems or their components must maintain a desirable level of service. The percentage of risk of physical injury or damage to health directly correlates with damage to property or the environment.

Users and firms of IIoT struggle to effectively manage asymmetries between multivariable system design implementation decisions and their consequential concerns. Resetting this asymmetry requires essential information disclosure as the major component, which will have limited usefulness without implementing the desired steps.

After applying the methodologies of IIoT security, organizations are still incapable of escaping hacking as part of the broader digitally networked terrain. There is a requirement to develop a sense of individual and collaborative responsibility toward the IIoT, establishing a general understanding considering all the measures to effectively respond and recover from hazards and pitfalls. Developing consistency, repudiation, and business confidence are the internal components and the areas where the focus should be on improving security. Learning from historical attacks by studying and analyzing them enables one to adapt and respond accordingly.

Insurance companies have launched various electrical safety action in the previous century to ensure safe products for both businesses and homes.

The ultimate objective of this protocol is to enhance the security of IIoT devices, strengthen user bonds, and improve end-user connectivity. Organizations focus on identifying potential threats and assessing risk exposure, with an overall objective of minimizing damage. IIoT safety must prioritize criteria that minimize risks. The objective is to utilize insurance programs, regulations, and structures to incentivize best practices in IIoT security design, implementation, and maintenance.

This protocol targets the following communities as its audience:

1. The finance sector, incorporating the insurance community industry.
2. Teams involved in deploying IIoT systems in production.
3. IIoT device and service manufacturers and producers.
4. International and national governance organizations are critically involved in protecting sensitive infrastructure.

The protocol supports the development of more spectacular expectations for insurance coverage claims. It improves the actuarial considerations in cybersecurity, addressing uncertainties in damage calculation. Industry-wide standards that are acceptable provide a basic framework to secure IIoT services effectively. Rather than waiting for government regulations, experts deploying the IIoT security systems should incorporate necessary incentives for adopting changes through self-regulation in operations and governance. This protocol aims to create a reliable production process for end users involved in manufacturing and production, ensuring a more trustworthy ecosystem for the related devices.

The improvement of safety and cybersecurity practices in IIoT system deployments will serve as a benchmark of expectations for other IoT deployments across the supply chain and may influence cybersecurity in consumer devices and smart cities. For governments, the Protocol provides a means of initiating a dialogue with domestic industries and its relation to concerns over the safety and security of critical infrastructure in the interconnected IoT environment. The Protocol supports mechanisms through which IIoT system providers can share information about their vulnerabilities in a way that maximizes safety and security in the public interest. In response to an IIoT incident, there is also an opportunity to enhance loss investigation, implement response strategies, provide emergency support, conduct IT forensics, offer specialist legal and public relations support, and provide funding support.

6.4.1 IIoT safety and security digital protocol

An entity should impose the mechanism of cybersecurity in case of an attack on its processes and products. The entity should design, develop, and implement these mechanisms to handle IIoT system and react throughout IT and OT environments. Governance should verify these strategies and risk management mechanisms must be in place. Throughout the entire life cycle, the entity must have the capabilities and the mechanism for analyzing, authorizing, and controlling IIoT security risks and vulnerabilities.

Toward this goal, the Protocol focuses on setting forth baseline requirements for insurability in three areas

1. Line of Business IIoT Device Safeguards
2. Internal Governance and Risk Management
3. Record Keeping and Data Management.

The entity deploying the IIoT system must incorporate key safeguards for the components of IIoT or the mechanisms it develops, creates, analyzes, interacts, and controls.

1. *Risk Assessment Models*: The risk assessment technique used by organizations installing IIoT systems must first identify all digital and physical assets that need to be safeguarded. Along with

thorough vulnerability analysis, the risk assessment model should identify the variables affecting IIoT system processes and potential threat agents.

2. *Hardware Integrity*: Hardware integrity must be guaranteed throughout the endpoint lifecycle due to changes in hardware configuration and components to prevent uncontrolled modifications to the hardware components. The theft of some hardware resources is a potential weakness of the device. The endpoint needs to have the ability to defend itself from unauthorized access and the monopolization of crucial resources like memory, processing cycles, and privileged processing modes.

3. *Encryption*: When adopting IIoT systems, organizations must ensure that applications and accompanying devices adhere to the latest universally accepted security and cryptographic requirements. All stored and transmitted personally identifiable information must be encrypted using the most recent generally accepted security standards.

4. *Patches and Streamlining*: To provide properly authorized software and/or firmware updates, patches, and revisions, entities adopting IIoT systems must have a framework for automatic safe and secure ways. Such updates need to be signed or otherwise confirmed to originate from a reliable source.

5. *Interoperability*: IIoT products and services must be able to exchange data using industry-standard protocols with each other and with base stations. For network traffic, devices should use standard ports.

6. *Software Development Lifecycle*: In addition to maintaining an inventory of the source for any third-party/open-source code and/or components used, entities deploying IIoT systems must ensure that all IIoT devices, services, and associated software have undergone a rigorous, standardized software development lifecycle process and methodologies, including unit, system, acceptance, regression testing, and threat modeling. These entities should adopt commonly known code and system hardening strategies to prevent data leaks between the devices, apps, and cloud services in various common use case scenarios and settings.

7. *Service Trusted Computing Base*: By standardizing the computing platform and defining the collection of apps, libraries, and configuration files that will run on the computing platform, entities providing IIoT services are required to create a Service Trusted Computing Base. They should establish a method for creating an application image, cryptographically signing it, and verifying it after signing.

8. *Organizational Root of Trust*: To ensure that every computer platform in the IIoT service is authenticated when communicating with other computing platforms, entities deploying IIoT systems must build a cryptographic-based system. They should create a root secret and/or certificate, ensuring their safekeeping and protection throughout their lifecycle.

9. *Vulnerability Exposures*: In order to receive, track, and effectively address complaints of external vulnerabilities from third parties, entities using IIoT systems must develop coordinated vulnerability disclosure processes and mechanisms. Programs like "bug bounty" and crowdsourcing should be taken into consideration by developers to help find vulnerabilities that internal security teams may not be able to find or detect.

6.4.2 Internal governance and risk management

An organization adopting IIoT systems must show that the IIoT systems it designs, develops, produces, installs, maintains, monitors, interacts with, or controls have acceptable internal governance and risk management measures. *Advancing Cyber Resilience: Principles and Tools for Boards*, published by the World Economic Forum, offers a business model and best practices for such mechanisms at the Board level.

1. *Board Oversight*: As part of the company's risk management strategy (avoidance, reduction, sharing, and retention) and business continuity plans, the entities deploying IIoT systems' board and senior leadership must formally review the organization's IIoT cyber strategy (prevention, transfer, and response), engage in governance, and oversee this strategy.
2. *Top Position Responsibility*: When installing IIoT systems, organizations must designate a "Responsible Officer" for cybersecurity and resilience and ensure that business and IT staff are properly versed on the topic. Entities installing IIoT systems must additionally have an officer in charge of organizational security/resilience and responsibility assignment matrix (RAM) implementation, in addition to, or as part of, this function.
3. *Cyber Resilience*: Employing a combined strategy for People, Capital, and Technology, entities implementing IIoT systems must show that cyber resilience is incorporated into business strategy, quantify and identify organizational cyber risk strategy, and determine organizational cyber risk assessment.
4. *Ongoing Assessment*: Assets throughout the service and endpoint ecosystems must be regularly and thoroughly examined by organizations using IIoT systems.
5. *Ongoing Testing*: In addition to preparing for and following IIoT security best practices throughout the IIoT service's distribution, installation, service, and maintenance channels, entities deploying IIoT systems must periodically test the IIoT service's cybersecurity and resiliency using penetration testing and other tried-and-true security methods.

6. *Track and Address Legacy Systems*: Organizations installing IIoT systems need to start processes to monitor and deal with legacy and out-of-date solutions and ensure they receive proper maintenance.

7. *Information Sharing*: The exchange of information regarding risks and vulnerabilities with trusted intermediaries from the business sector or government agencies must be operationalized by entities adopting IIoT systems.

Security contacts each component of an IIoT gadget and framework lifecycle; therefore, IIoT shields require cross-functional, cross-departmental, and cross-organizational coordinated effort to accomplish. as the following are significant protection considerations for planning, producing, administration, circulation, joining, and different purposes of IIoT, making an Obligation Task Network for staff to carry out dynamic security solidifying:

- Determining the processes, systems, and devices that make up its IIoT exposure.
- Assessment of security vulnerabilities and a plan for closing gaps.
- Assessment of the secure configuration and a plan for closing the gap.
- Plan for closing the security hole in the application.
- Plans for closing gaps, patch assessments, and secure management.
- Plans for closing gaps and securing data storage and transport.
- End-of-life assessments and a plan for remediation, including secure firmware, software, hardware, and application upgrades.
- During the design, commissioning, and RUN phases, as well as gap remediation plans, secure integration testing, penetration testing, and compliance testing are performed.

6.4.3 Record keeping and metrics

Reports on the security of an association's IIoT systems should be covered by decision-makers from the inception of the system's general use through its operation. Drivers, stakeholders, and other decision-makers are informed by the applicable criteria and measures. While certain operations will differ depending on particular situations in which they are being used, others are also very effective in detecting tried attacks, and the exact flow of the protocols related to the security that have urged a disquisition.

Pointers that indicate efforts, security metrics like data sources, communications, and overall performance identifiers must be easily and directly represented in dashboards and other visualizations by associations deploying IIoT systems. This will permit functional and business personnel to make better business decisions. As a result, safety emerges as a valuable element of the functional process, with its components reflected in the costs avoided by preventing bad opinions.

In order to ensure a nonstop feedback circle, associations installing IIoT systems must introduce safety parameters to provide quantifiable inputs for effective decision-making, improve responsibility, and demonstrate compliance with all the legal aspects. These criteria aid in faster and more effective operation and governance by aiding in the early discovery of security issues.

6.4.4 Operation of protocol

6.4.4.1 Assessment mechanisms

This convention no doubt serves as a good boosting system. IIoT patron might refuse to give protection to an IIoT establishment except if necessities for IIoT uncertainty in the Member are met. The following are likely pointers of insurability for IIoT insurers.

Internal security measures applicable to ensure that IIoT company adheres to Part 1 and views safety as an essential component of its entire business prospect, thereby ensuring compliance with Section 1's conditions, including applicable IIoT norms. It's likely that the IIoT insurer will choose the applicable standard for each use.

The assurance of the entire appreciation of the security of IIoT, its insurers may contribute applicable information and analysis to the pointers and data institute. Means are sufficient to maintain and modernize IIoT systems that have formerly been stationed (also referred to as "heritage" systems), agreement with Portion1 to guarantee security against evolving IIoT security pitfalls throughout the device and system's life cycle (21).

6.4.5 Insurability conditions

The honest operation of the IIoT insurer's terms will impact reinsurance assessments. The conditions outlined in Sections 6.1, 6.2 and 6.3 Portion 1 encapsulate the evaluation for establishing IIoT insurability. This evaluation will be based on the entire risk factors, components, and perpetration. IIoT insurers will specify the particular conditions for insurability.

6.4.6 Perpetration of protocol

1. *Safeguards Assurance*

    The mechanism will be implemented as soon as insurability is determined, if not earlier. Systems or devices from the IIoT's history will not be grandfathered in prior to Draft – For Discussion Only 11 for granting new insurance or renewal of existing policy. Section 1b and other conditions of this document will be used by IIoT insurers to estimate IIoT enterprises. The insurer will not issue any insurance or assume any other risk for the IIoT organizations until they comply with Section I's conditions, thereby, providing pivotal incentives for IIoT security.

2. *Pointers and Data Clearinghouse*

When determining the insurability of a system or an IIoT company, data on safety compromises and occurrences involving the components or implementations are essential. This protocol recommends forming an institute for IIoT enterprises and insurance vendors to incorporate this data and assess insurability factors and risk analysis to guarantee their vacuity. IIoT companies and insurers will give data and pointers to this institute. An expert network will develop and operate protocols for the institute. The pointers and data outlined in Section I C will accept applicable data and pointers from IIoT enterprises, insurers, and interested third parties (security vendors, expert advisors, and governing bodies.). This institute serves as a necessary wellspring of the data important to estimate the insurability of the IIoT terrain. Insurance for IIoT companies that do not comply with or violate the protocol will be suspended until the issue is fixed.

3. *Verification of Protocol*

The verification mechanisms of this protocol are related to insurers and IIoT businesses. The effectiveness of the Protocol's factors and this framework's capability to incentivize security through insurance will be assessed by IIoT businesses and insurers. The IIoT community will determine and regularly test and estimate verification procedures.

4. *Conservation of Protocol*

This protocol will address the latest trends, security parameters, cybersecurity principles, and extraordinary efforts from time to time to maintain its connection in light of evolving security precautions. To guarantee that the standards already available in Section 1 remain current and that associations deploying IIoT systems continue to adhere to applicable norms for both new and existing IIoT setups, IIoT insurers will regularly conduct surveys to assess the coverage of IIoT security norms. As this protocol governs the deployment of IIoT systems and insurance, practices, and updated conservation procedures will be implemented. IoT must be integrated within the entire farmwork as it represents a product of digital development and forms the cybersecurity infrastructure, the mode in which the involved components will communicate. Network members should monitor the developments in space, machine literacy, robotics, and amount computing to ensure the protocol remains relevant and effective. This aspect is pivotal when considering the role of insurance brokerage and how to deal with the multitudinous IIoT products, systems, and cyberinfrastructure. The affected community will determine conflict resolution arising from this protocol. Any medium for resolving a conflict must be open and provide all parties involved with the chance to present their arguments to a neutral third party.

REFERENCES

1. R. H. Weber, "Internet of Things – New security and privacy challenges," *Computer Law & Security Review*, vol. 26, pp. 23–30, 2010.
2. N. Koshizuka and K. Sakamura, "Ubiquitous ID: Standards for ubiquitous computing and the Internet of Things," *IEEE Pervasive Computers*, vol. 9, no. 4, pp. 98–101, Oct.–Dec. 2010.
3. N. Kushalnagar, G. Montenegro, and C. Schumacher, "IPv6 over Low Power Wireless Personal Area Networks (6LoWPANs): Overview, assumptions, problem statement, and goals," *Internet Eng. Task Force (IETF), Fremont, CA, USA*, RFC4919, vol. 10, Aug. 2007.
4. G. Montenegro, N. Kushalnagar, J. Hui, and D. Culler, "Transmission of IPv6 packets over IEEE 802.15. 4 networks," *Internet Eng. Task Force (IETF), Fremont, CA, USA*, Internet Proposed Std. RFC 4944, 2007.
5. S. S. Saini, S. S. Malhi, B. K. Saro, M. S. Khan, and A. Kaur, "Virtual reality: A survey of enabling technologies and its applications in IoT," *2022 International Conference on Electronics and Renewable Systems (ICEARS)*, Tuticorin, India, 2022, pp. 597–603, https://doi.org/10.1109/ICEARS53579.2022.9752424.
6. Technical report by N.Srivastava on RFID: RFID Introduction, Present and Future applications and Security Implications, George Mason University, Fairfax, VA, Fall, 2006.
7. A. Kaur, G. Singh, V. Kukreja, S. Sharma, and S Singh. "Adaptation of IoT with blockchain in food supply chain management: An analysis-based review in development, benefits and potential applications," *Sensors*, vol. 22, no. 21, pp. 8174, 2022.
8. V. Sharma, N. Mishra, V. Kukreja, A. Alkhayyat, and A. A. Elngar, "Framework for evaluating ethics in AI," In *2023 International Conference on Innovative Data Communication Technologies and Application (ICIDCA)*, 2023, pp. 307–312).
9. G. V. Crosby and F. Vafa, "Wireless sensor networks and LTE-A network convergence," In *Proceedings 38th Annual IEEE Conference on Local Computer Networks*, 2013, pp. 731–734.
10. A. Dunkels, B. Gronvall, and T. Voigt, "Contiki-A lightweight and flexible operating system for tiny networked sensors," In *Proceedings. 29th Annual IEEE International Conference on Local Computer Networks*, 2004, pp. 455–462.
11. P. Barnaghi, W. Wang, C. Henson, and K. Taylor, "Semantics for the Internet of Things: Early progress and back to the future," *Proceeding of IJSWIS*, vol. 8, no. 1, pp. 1–21, Jan. 2012.
12. T. Kamiya and J. Schneider, "Efficient XML Interchange (EXI) Format 1.0," World Wide Web Consortium, Cambridge, MA, Recommend. REC-Exi-20110310, 2011.
13. M. Sun, Y. Liu, and K. Liu, "Security problem analysis and security mechanism research in IoT," *Journal of Secrecy Science and Technology*, vol. 18, no. 3, pp. 1–14, Nov. 2011.

Chapter 7

Data breaches in sensors through various integration techniques

Shabdapurush Poudel and Sudan Jha

7.1 INTRODUCTION

In recent times, we have seen a phenomenal increase in the usage of industrial-grade IoT devices in our day-to-day lives. We can say the Industrial Internet of Things (IIoT) represents the fourth industrial revolution, with the potential to impact industries on a scale equal to prior advancements such as steam, electrical, and nuclear technologies. IIoT has brought about various advancements in the way industries operate and has significantly changed how data is collected, analyzed, and utilized. However, with the rapid growth of IIoT devices, we have also seen a rise in cyber-attacks and physical attacks targeting IIoT, which are sophisticated enough to exploit vulnerabilities from individual components to the systems level.

To mitigate these risks, it is very important to set up security standards that can ensure the safe use of IIoT devices, from ensuring confidentiality, integrity, and availability of data collected and transmitted between these devices [1]. Security standards are the backbone to implementing such security measures, policies, and procedures and help the industry identify such vulnerabilities and address the issue in time to prevent loss.

7.1.1 Definition of IIoT security standards

IIoT security standards refer to a set of guidelines, best practices, and protocols designed to secure the IIoT system devices from unknown cyber-attacks, breaches, and other security threats. IIoT security standards can help create guidelines for a secure and reliable environment for devices and networks by addressing the unique security challenges and risks involved [1]. Compliance with these security guidelines and standards can help industries meet the requirements to reduce the risk of cyber incidents and protect data.

7.1.2 Importance of IIoT security standards

IIoT security is crucial in ensuring the safe and reliable operation of systems and processes. With the increase in the adoption of connected

DOI: 10.1201/9781003530572-7

devices, the risk of cyber threats and breaches has also increased. These threats can result in significant financial losses, operational downtime, and even physical harm.

IIoT security standards aim to protect IIoT devices and networks from incoming threats by identifying and monitoring risks and helping to fix vulnerabilities. By adopting security-by-design principles and adhering to these security standards, industries can minimize the risk of cyber-attacks and ensure the safe operation of their systems [2].

Proper integration of the IIoT can optimize the use of assets, predict points of failure, and even trigger maintenance processes autonomously. This can result in improved efficiency, reliability, and cost savings for industries. However, without proper security measures in place, these benefits can be quickly negated by the risks posed by cyber threats [3].

In summary, IIoT security is essential for ensuring the safe and reliable operation of industrial systems and processes. By following the established guidelines, industries can minimize the risk of cyber threats and maximize the benefits of the IIoT.

7.1.3 Overview of IIoT security standards

As countermeasures to security breaches and cyber-attacks, various guidelines and standards have been developed. The *NIST Cybersecurity Framework* was developed by the National Institute of Standards and Technology (NIST), which introduces a set of guidelines for industries and organizations to manage and reduce cyber-attacks and other threats [4]. *ISO/IEC 27001* is a set of standards in the IT sector that provides requirements for an Information Security Management System (ISMS), representing a systematic approach to managing sensitive company information and reducing and managing risks [5]. Similar to ISO 27001, *IEC 62443*, developed by the International Electrotechnical Commission (IEC), includes a set of standards and technical reports that deal with industrial cybersecurity. It was designed to help end users, system integrators, and manufacturers reduce the risk of deploying and operating an IACS [6]. Along with *IEC-62443*, *ISA/IEC-62443* (ISA) also provides a set of recommended standards that help companies with industrial automation and control systems protect and secure those systems. Another standard, *UL 2900*, developed by Underwriters Laboratories (UL) provides a series of standards for cybersecurity, addressing the testing and certification requirements for products, processes, as well as specific industry and network systems [7].

Although not as automation and technically oriented, the General *Data Protection Regulation (GDPR)* is a European Union (EU) regulation that outlines the requirements for the protection of the personal data of EU citizens. In the IIoT scenario, implementing a GDPR policy can make it difficult to gain the consent that is needed to process personal data within connected networks [8]. A few other non-technical *HIPPA* security standards aim to protect the medical data of individuals and provide a set of guidelines, for medical device manufacturers to implement security-by-design for products, making the devices immune to severe cyber-attacks.

These IIoT security standards provide frameworks and guidelines for securing IIoT system devices and can be used in conjunction with one another to improve security. By implementing these standards, industries can identify potential vulnerabilities and loopholes, and implement appropriate security measures to protect their IIoT systems from the development phase itself.

7.2 COMMON IIOT SECURITY THREATS

7.2.1 Cyber-attacks

Cyber-attacks are the most common form of security threat in IIoT. Data siphoning, device hijacking, distributed denial-of-service (DDoS), device spoofing attacks, physical device theft, and data breaches through legacy systems are some of the examples of security risks.

Various examples of cyber-attacks in the IoT context include:

> *Mirai Botnet: In October 2016, Internet performance management services provider Dyn was hit with the worst DDoS IoT botnet attacks. Using well-known default usernames and passwords, the attackers logged in and continuously scanned the web for susceptible IoT devices to infect them with malware* [9].
> *Verkada Attack: A cloud-based video surveillance service, Verkada, March 2021, was attacked where the attackers accessed over 150,000 live feeds of cameras by obtaining and using legitimate admin account credentials* [9].
> *Jeep Hack: In July 2015, a group of researchers took control of the Jeep SUV vehicle via a cellular network, controlling the vehicle's speed and steering of the vehicle while testing* [9].

Industries need to be aware of these cyber-attacks and take appropriate measures against them. This can include following established security standards, implementing strong access controls, and regularly updating and patching systems.

7.2.2 Malware and ransomware

Malware and ransomware are other common forms of threats to IIoT. Although IIoT devices rarely store critical data locally, this doesn't mean they are protected from ransomware attacks. Ransomware attacks in IIoT generally target the devices and stop their core functionality instead of ransoming data. Ransomware generally stops the devices from functioning normally and can control the entire system. This can mean shutting down fundamental operations, having a major impact on operations, and forcing payment under pressure.

Malware can be used as a middleman for extracting data from the IIoT devices by hiding itself as a core program in the IIoT component software and having control over the device. *The Wannacry Attack in May 2017* targeted devices operating on Microsoft Windows and encrypted all the data in the computer, demanding ransom for unlocking Bitcoin cryptocurrency [10]. *Flocker* started with an Android mobile lock screen and gradually shifted to TVs and thermostats of the house, assuming complete control over them [11].

To prevent malware and ransomware attacks, industries and organizations should regularly push patches and updated security countermeasures. Continuously monitoring for any anomalies in the network system and application logs and restricting access to any unauthorized devices, emails, and scripts can also help prevent malware and ransomware threats.

7.2.3 Physical security threats

Physical security threats involve the physical manipulation, theft, or damage of IIoT devices or systems, which can compromise the security and integrity of the data being collected and transmitted. Here are some examples of physical security threats in IIoT: tampering, theft, physical damage, and unauthorized physical access to IIoT devices.

An attacker gaining access to a controller that operates a water treatment plant, allowing them to manipulate the system and potentially cause harm to the environment or public health, can be considered a scenario of physical security threat.

Stuxnet Attack: The Stuxnet IoT attack, is a well-known IoT attack that targeted a uranium enrichment plant in Natanz, Iran. The attackers targeted the Siemens Step7 software running on Windows via a USB controller and compromised the system, giving the worm access to the industrial program logic controllers. This allowed the attackers to control different machines at the industrial sites and access vital industrial information [9].

To prevent physical security threats in IIoT, organizations should implement appropriate physical security measures, such as access control, surveillance, and tampering detection mechanisms. Additionally, organizations and industries should regularly ensure that IIoT devices are installed in secure locations and routinely inspected for signs of tampering or physical damage.

7.2.4 Insider threats

Insider threats can be a major risk to an organization's data integrity and classified information. A person with authorized access to the company's credentials can pose a severe risk, as they may leak information or cause significant problems to the organization by compromising confidentiality due to malicious intentions.

One example of an insider threat in IIoT is the case of a former employee of an American water treatment plant who used his knowledge of the plant's control systems to remotely access them and change the levels of chemicals used to treat the water. This could have resulted in significant harm to the public if it had not been detected and corrected in time [12].

It is important for industries to be aware of these kinds of risks and take appropriate measures such as profiling staff, implementing strong access controls, regularly monitoring user activity logs, and providing training on proper security procedures. Periodically changing the authentication method can also help prevent these types of threats.

7.3 DATA BREACH IN IIOT

Data breaches are recurring threats that plague IIoT devices, as the process of data breaches becomes increasingly complicated and sophisticated. This risk involves attackers using an IIoT device as a doorway to the central network and gaining access to critical data [13]. Data breaches occur when sensitive or confidential data is disclosed without permission, potentially resulting in significant financial losses, damage to reputation, and even physical harm. Data breaches usually target client or partner data, personally identifiable information, health data, and financial data due to a lack of encryption for critical data (Figure 7.1).

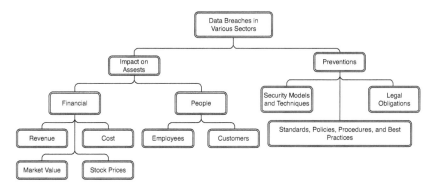

Figure 7.1 Data breach taxonomy.

7.3.1 Common causes of data breach

Data breaches can occur due to various causes, among which some of the most common are:

Weak Password Protection: Weak password combinations and hard-coded credentials are often the culprits behind data breaches. The use of easily guessable credentials, default credentials, and neglecting periodic credential changes can be exploited by attackers, leading to severe data breaches.

Lack of Regular Patches and Weak Update Mechanism: Many IIoT devices are not patched regularly due to their embedded system nature. Additionally, many devices do not even receive regular updates after leaving the production assembly, leaving vulnerabilities and threats that can lead to open loopholes for attackers to exploit and gain access to critical data.

Insecure Interfaces: Insecure open interfaces that IIoT devices connect to, and the APIs they use, can also serve as sources for data breaches.

Unpatched Applications: Many IIoT devices run on open-source software that is customized by industries for their intended use. Most software on these devices is left unchecked for many years. If these applications are left unpatched and unchecked for regular threats, they can serve as doorways for data breaches.

Here are a few examples of data breach scenarios that have already occurred:

- During the Target data breach in 2013, hackers gained access to Target's network through an HVAC vendor and stole credit and debit card information from 40 million customers.
- An Amazon-owned company, Ring, accidentally revealed several users' data to both Facebook and Google via third-party trackers in their Android application during a previous data breach [14].
- In another incident, Ring experienced an IoT breach where hackers spied on multiple families via the Ring Doorbell and connected cameras [14].

7.3.2 Data loss prevention

Data Loss Prevention (DLP) ensures that data is not lost, misused, or accessed by unauthorized personnel using established guidelines and processes. DLP software classifies confidential data driven by regulatory compliance such as Health Insurance Portability and Accountability Act (HIPAA), Payment Card Industry Data Security Standard (PCI-DSS) or General Data Protection Regulation (GDPR). Once data is identified, DLP employs remediation techniques such as multiple encryption algorithms, multi-factor authentication, and other methods to protect against breaches. DLP also scans each node and endpoint for vulnerabilities and provides insights for enhanced protection.

DLP addresses three main objectives of data policy for many organizations [15]:

- *Personal Information Protection/Compliance*: Data protection isn't a case of whether the organizations want it or not, but a requirement to adhere to regulations such as HIPAA, PCI-DSS, and GDPR.
- *Protection of Intellectual Property*: An organization's IP, trade policies, and trade secrets are critical and should be protected from external risk factors.
- *Data Visibility*: How data is stored and transmitted between endpoints should also be a matter of importance for organizations, hence data visibility.

DLP should be of utmost priority for all organizations and organizations can follow some simple steps, such as [16]:

Secure Database Access: Securing the database and hardening it against cyber-attacks can prevent data loss via protection against SQL Injection. This can be achieved by having granular access controls for the database and establishing a separate, secured node for database communication [16].

Sensitive Data Encryption: The goal of encryption is to enhance security processes such as authentication, authorization, integrity, and non-repudiation, making it an effective method for data protection. Proper implementation of strong encryption algorithms can significantly enhance the protection of organizations [16].

Fine-Grained Access Controls: Along with the use of identity management, fine-grained access control is also critical to prevent data loss. Fine-grained access control can provide access to certain parts of the program, process, or data using a secured channel and proper authentication. This can also help invoke or revoke privileges at a high level to restrict data loss. Enforcing such a high level of security measures can be challenging for an organization, though organizations can move toward no-default credentials and granular steps of access controls [16].

Data Backups: Maintaining regular backups of critical data flagged by regulations in different locations and scenarios creates an optimal environment for ensure data protection. Additionally, regular tests should be conducted to ensure that the backed-up data is not corrupted or susceptible to data breach threats [16].

Patches and Software Updates: Patches should be a crucial element of any business's data loss prevention strategy. Regularly patching software against easily exploitable vulnerabilities can prevent data leak, and loss. Also, updating software on a regular basis ensures that the software has as

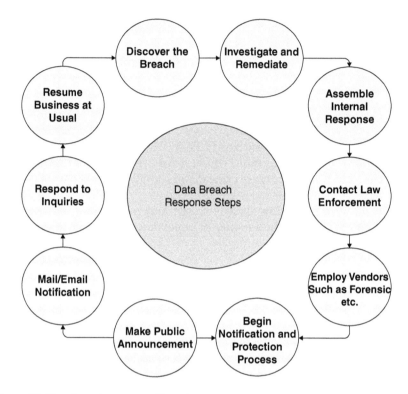

Figure 7.2 Data breach response lifecycle.

few loopholes as possible and reduces severe risks [16]. Software updates not only include patches but can also introduce new features to ensure data loss prevention and efficiency.

The data breach response lifecycle is a way to understand how responses to certain data breaches are handled, consisting of parts that include identifying, notifying, and resuming business as usual (Figure 7.2).

7.4   IIOT SECURITY STANDARDS

7.4.1   NIST Cybersecurity Framework

This framework, created by the National Institute of Standards and Technology, offers guidance on how to respond to, avoid, and recover from cyber incidents, helping to create better cybersecurity. The NIST Cybersecurity Framework defines rules and best practices that can assist businesses in developing and strengthening their cybersecurity position [4,17].

The core functions of the NIST Cybersecurity Framework are organized into five key points, which are:

Identify Function: This function focuses on laying the groundwork for an effective cybersecurity program by
- defining the organization's business environment and physical and software assets to form the framework of an asset management program [17],
- identifying established cybersecurity policies and legal and regulatory requirements related to cybersecurity capabilities,
- determining asset vulnerabilities and threats to internal and external organizational resources,
- identifying risk tolerance by establishing a risk management strategy,
- identifying a supply chain risk management strategy including priorities, constraints, risk tolerance, and assumptions to support risk decisions [4].

Protect Function: This function outlines appropriate safeguards to ensure delivery and support the ability to contain potential cybersecurity events by
- implementing protections for Identity Management and Access Control, both physical and remote access,
- empowering staff through security awareness training,
- implementing procedures to maintain and manage the protection of systems and assets,
- protecting organizational resources through maintenance,
- managing technology to ensure the security and resilience of systems [4,17].

Detect Function: This function defines the appropriate activities to identify the occurrence of cybersecurity events in a timely manner by
- ensuring anomalies and events are detected and studying their potential impact,
- putting in place capabilities for continuous monitoring to track cybersecurity incidents and assess the efficacy of safety precautions [4,17].

Respond Function: This function focuses on appropriate activities to take action in case of a detected incident by
- managing interactions with both internal and external stakeholders during and after an occurrence, and ensuring that the reaction planning process is carried out [17],
- performing rectification and mitigation to prevent loss and resolve [4].

Recovery Function: This function identifies activities to renew and maintain plans and restore any services that were hampered during the incident by

- ensuring the organization implements recovery plans and procedures to restore the system and implement improvements based on experience,
- coordinating between internal and external communications during the recovery from an incident [4].

7.4.2 ISO/IEC 27001

ISO/IEC 27001 is the best-known security standard in the IT sector as it provides requirements for an Information Security Management System (ISMS), which is a systematic approach to managing sensitive company information, reducing, and managing risks. Risk Management is the foundational standard and is based on processes that help organizations identify, assess, and manage information security risks.

The standards are relevant to IIoT because they provide a comprehensive framework for securing both IT and OT systems. Organizations can utilize flexibility and tailor the requirements to their specific needs. Some benefits of implementing these standards include [5,6]:

- *Risk Management*: ISO/IEC 27001 can provide a structured approach for identifying and assessing information security risks of IIoT systems. By implementing standards, organizations can actively address the risks and reduce the risks of breach.
- *Compliance*: Implementing ISO/IEC 27001 can help organizations accept and implement regulations that evolve along with the IIoT space. It can also help organizations demonstrate compliance with industry-specific obligations such as the Network and Information Systems (NIS) and General Data Protection Regulation (GDPR).
- *Improved Operational Efficiency*: By implementing ISO/IEC 27001, organizations can streamline their security processes. This can lead to increased efficiency and cost reduction, as an effective ISMS can help detect and prevent incidents, minimize downtime, and protect critical assets, thereby increasing efficiency.
- *Better Reputation and Trust*: Having ISO/IEC 27001 certification can demonstrate to customers and clients that the organization takes security as a concern and is committed to maintaining a high level of information security to protect sensitive data.

However, implementing the standard is not an easy task as it has some challenges too, such as:

- The most obvious and big challenge can be the complexity of the IIoT system itself. The interconnected nature of the system can help detect and identify potential risks and appropriate solutions.

- Since IIoT devices also have to work in harsh environments, it can be challenging to implement physical security controls.
- Since the cost of implementing and maintaining an ISMS can be high, this can pose a challenge for small and medium-sized organizations.

Although there are some challenges in implementing ISO/IEC 27001 standards, they can serve as a framework for implementing and maintaining effective ISMS for IIoT systems. This can help organizations by reducing the likelihood of security incidents and transforming organizational manufacturing efficiency.

7.4.3 IEC 62443

Developed by the International Electrotechnical Commission (IEC), IEC 62443 is a comprehensive set of security standards designed for Industrial Automation and Control Systems (IACS). Its objective is to help end users, system integrators, and manufacturers reduce the risk of deploying and operating an IACS. This standard is designed to be flexible and adaptable to different IACS systems and can be utilized throughout various industries' life-cycles, including manufacturing, energy, and transportation. It includes nine standards, technical reports, and technical specifications. IEC 62443 adopts a risk-based approach to cybersecurity, based on the idea that trying to safe-guard all assets equally is neither effective nor sustainable. Instead, users must prioritize what is most valuable and needs the most security, as well as pinpoint weaknesses. [6,18–20]. The IEC 62443 is organized into four parts:

- *Terminology, Concepts, and Models*: This part provides definitions for terms and concepts related to IACS security and also mentions the various security models to secure IACS systems.
- *Policies and Procedures*: This part guides on developing security policies and procedures for IACS systems and covers topics such as risk assessment, security planning, and security awareness.
- *System Requirements*: This part covers topics such as access control, network segmentation, and intrusion detection and also outlines the security requirements that IACS systems should meet.
- *Component Requirements*: This part provides guidance on security requirements for individual components of IACS systems and also includes communication protocols and interfaces.

The benefits of IEC 62443 implementation are:

- The standard provides a holistic approach to security and covers all the security aspects, including risk assessment, security controls, and continuous improvements.

- The standard also emphasizes the importance of a defense-in-depth approach to security, which means that IIoT systems can have multiple layers of security controls in place to protect against both internal and external threats.
- Another important aspect of IEC 62443 is that it provides guidance on secure system design and includes recommendations for secure coding practices, secure hardware and software design, and secure configuration management.

In conclusion, the IEC 62443 security standard provides an extensive framework for the implementation and maintenance of secure IIoT systems. By implementing this security standard, companies can secure their IACS systems and increase automation efficiency.

7.4.4 UL 2900

These cybersecurity standards, created by Underwriters Laboratories (UL), address the testing and certification requirements for goods and services as well as network and industry-specific systems, and they help increase the security of vitally connected electronic physical security systems. The UL 2900 is divided into several sections that offer instructions for various facets of cybersecurity testing and certification [7].

- *General Requirements:* This part provides an overview of the testing and certification process, as well as guidelines for scope and objectives.
- *Technical Specifications*: This part provides the technical specifications for testing network-connected products and systems, covering topics such as vulnerability testing, penetration testing, and security control testing.
- *Cybersecurity Criteria*: This part provides a set of criteria for assessing the cybersecurity of network-connected devices and systems, covering topics such as access control, data protection, and incident response.
- *Process Requirements*: This part provides guidelines for the testing and certification process, including requirements and reporting.

Key benefits of implementation of UL 2900 are:

- This standard provides a comprehensive approach to cybersecurity testing, covering a wide range of cybersecurity risks, from malware and unauthorized access to data breaches and physical attacks.
- It also emphasizes the importance of ongoing testing and certification, which means that IIoT systems should be regularly tested to remain secure even in the face of evolving threats.

- UL 2900 has a three-tier security approach, with each tier increasing the level of security. Security tests include fuzz testing, known vulnerability detection, code and binary analysis, risk control analysis, structured penetration testing, and security risk control assessment.

In conclusion, UL 2900 provides an extensive framework for cybersecurity testing and certification. It can also help organizations by providing the benefits of security testing to test the integrity of the devices and assess the security of interconnected device systems.

7.4.5 GDPR

The General Data Protection Regulation (GDPR) is a data protection regulation introduced by the European Union in 2018. Although focused mostly on protecting personal data, it can also serve as a security standard for IIoT systems. The GDPR requires organizations to implement technical measures to ensure the security of personal data, which also includes protecting against unauthorized access, disclosure, alteration, and destruction of critical data.

The GDPR itself doesn't provide specific guidelines for security; instead, it offers a comprehensive framework for organizations to follow when implementing security measures.

The benefits of using GDPR as security standards are:

- GDPR emphasizes the importance of risk assessment, requiring organizations to conduct regular risk assessments, identify potential data security risks, and implement countermeasures.
- GDPR also requires organizations to implement technical measures for data security, such as high-level encryption, multi-factor authentication, and access control.
- GDPR requires organizations to implement appropriate measures to ensure data confidentiality, integrity, and availability of data. This includes using measures such as the DC-DR concept (Data Center – Data Recovery) and incident response.

Some challenges are also bound to arise while implementing the GDPR as security standards:

- *Complexity*: Implementing advanced security for data protection can also introduce complexity in coding and designing systems.
- *Lack of Clarity*: Since GDPR defines personal data broadly, this can create confusion among organizations, as to categorizing data as critical and identifying measures to protect them.

- *Technical Difficulties*: Implementing GDPR requirements can lead to major changes in technical design, as IIoT systems may need to be retrofitted or redesigned to comply with the requirements.
- *Regulatory Compliance*: GDPR compliance requires ongoing monitoring and reporting, necessitating IIoT systems to navigate complex regulatory frameworks and ensure data compliance.

In conclusion, GDPR is a regulation that provides guidelines on the collection and processing of critical and personal data. While GDPR implementation can present a few challenges, it serves as a guideline for protecting critical data. Considering the severe cases of data breaches, implementing GDPR can bring about security changes in IIoT systems, which can help protect individuals' data in the long run.

7.5 IMPLEMENTATION OF SECURITY STANDARDS

Implementing IIoT security standards requires an approach involving several steps. Some general steps that organizations can adhere to implement security standards are discussed below.

7.5.1 Risk assessment

Risk assessment is the process of identifying, analyzing, and evaluating potential vulnerabilities in IIoT systems. Conducting a risk assessment can help organizations identify weak areas in the system and measures to rectify these problems. The process includes several steps, such as [21]:

- Identify assets.
- Identify threats.
- Assess vulnerabilities.
- Determine the likelihood of risks.
- Determine potential impact.
- Develop mitigation strategies.
- Monitor and review.

Organizations must conduct risk assessment periodically, or in the case of

- Significant system changes.
- Security incidents.
- Regulatory changes.
- Periodic review.

Maintaining a regular risk assessment log and frequency can help organizations identify potential risks and patch them in time to prevent cyber risks, helping organizations protect against data breaches.

7.5.2 Security policies and procedures

Security policies and procedures provide guidelines to both employees and organizations on how to protect IIoT systems and data, ensuring that security controls are enforced. They help strengthen security by implementing rigorous procedures.

Implementing security policies and procedures involves key steps, such as:

- Identify applicable standards.
- Develop a security policy.
- Establish security procedures.
- Communicate policies and procedures.
- Monitor compliance.
- Update policies and procedures periodically.

Security policies and procedures define the requirements and methods for managing security in IACS environments and addressing the challenges. Standards such as ISA/IEC 62443 provide guidelines for managing these challenges by outlining security requirements at each level.

7.5.3 Access control

Access control is one of the most critical components in implementing security standards. It refers to the methods and techniques used to restrict access and use of IIoT systems and collected data to authorized personnel only. It helps control of data against unauthorized access and modifications and is also crucial in mitigating data breach scenarios. The implementation of access control involves several key steps:

- Identify access requirements.
- Implement authentication.
- Implement authorization.
- Implement audit trails and log trails.
- Implement physical access control.
- Review and update access control.

Access control can conduct privilege analysis to determine which user needs what access and can implement multi-factor authentication mechanisms to ensure only valid logins can get access to the system and data. It can be role-based or attribute-based access, limiting the access of intruders to protect critical information.

7.5.4 Incidence response

Incidence response refers to the process of detecting, identifying, and responding to security threats and incidents in IIoT systems, such as cyber-attacks, data breaches, or other physical incidences. Quick response to the problems arising in the IIoT system can be critical in preventing huge losses. Some steps involved in the implementation of incidence response are:

- Developing an incident response plan.
- Training employees.
- Establishing incident response teams.
- Establishing secure communication channels.
- Implementing incidence detection and monitoring.
- Conducting post-incident analysis.

Creating a team of experts who are properly trained to quickly respond to incidents of security breaches and threats is essential for implementing incident response as security standards. Also, establishing backup and recovery plans for handling threats and having proper plans and internal regulation guidelines for internal threats, equipment failures, and physical security breaches, can be considered within the scope of incident response. An effective incident response plan for IIoT systems can minimize the impact of security threats and data breaches, ensuring regulatory compliance.

7.6 INDUSTRY EXAMPLES OF IIOT SECURITY STANDARDS

Many industries have started implementing IIoT in their production and automation process. Some industries that have implemented IIoT along with standard security standards are discussed below.

7.6.1 Energy and utilities industry

The North American Electric Reliability Corporation Critical Infrastructure Protection (NERC CIP) standards, a set of requirements created to secure the North American bulk electric system, are used by the energy and utilities industries. This standard guarantees the reliable operation of the electric grid. Additionally, they follow ISO 27001 and the NIST Cybersecurity Framework [22].

Example: In 2015, a cyber-attack by Russian hackers on the Ukrainian power grid caused a blackout that affected over 225,000 customers. This attack showed the importance of implementing security controls, such as incidence response and security awareness training [21,23]. Energy and Utility companies implemented NERC CIP, NIST Cybersecurity, and ISO 27001 to minimize the risk of cyber incidents and secure their infrastructure from further cyber-attacks and distribution abruption.

7.6.2 Healthcare industry

The healthcare industry uses various security standards such as Health Insurance Portability and Accountability Act (HIPAA), NIST Cybersecurity Framework, and the Food and Drug Administration (FDA) regulations to protect critical data and mitigate the risk of unauthorized access to medical and drug equipment. These guidelines provide recommendations to manufacturers of medical devices to ensure they are resilient to cyber threats.

Example: In 2017, WannaCry ransomware attacked the United Kingdom's National Health Service (NHS), exploiting the vulnerabilities in Microsoft Windows OS. Various critical patient data were locked under ransom demands. This attack highlighted the severe threats of malware and ransomware on IIoT devices, emphasizing the importance of regular software patching and upgrades [23].

7.6.3 Transportation industry

The transportation industry uses several security standards for their IIoT-based system devices, such as the NIST Framework, IEC 62443, ISO 27001, and ISO 21177, which provide technical specifications addressing the securing of communication, signaling, and processing systems in railway applications. These standards provide guidelines for ensuring reliable and safe operation of transportation facilities [4–6].

Example: Russian cyber terrorists, known as EvilCorp, attacked GPS manufacturer Garmin on July 23, 2020. The attack compromised numerous crucial aviation systems, including navigation, autopilots, active traffic systems, flight instruments, engine information systems, displays, sensors, and interfaces. It also affected several marine equipment systems, including radars, chart plotters, sonar black boxes, automatic identification systems (AIS) sensors, and autopilots. Both GPS service providers and users of GPS products and services are required to protect GPS systems end-to-end [24].

7.6.4 Manufacturing industry

The manufacturing industry uses ISA/IEC 62443 series of standards to protect itself from cyber incidents in its IACS environment. These standards provide a framework for securing both the IT and OT systems in manufacturing environments and the importance of access control and network segmentation [25].

Example: In 2017, the NotPetya Malware attack disrupted Maersk's operations for weeks, causing an estimated $300 million in damages. This attack highlighted the importance of network-related security standards and automation protection security standards in the manufacturing industry [23].

7.6.5 Agriculture industry

The agriculture industry has started using IIoT in crop production, safety checks, and high-yield turnover processes to increase efficiency. The agriculture industry uses sets of standards over each other, such as ISO 27001, NIST Framework, and ISA/IEC 62443 to protect their connected network of agriculture devices [4–6].

Example: In 2018, a cyber-attack on John Deere (Agriculture Tool Manufacturing Company) by Fancy Bear, exploited their JDLink system to gain access and remotely control their equipment and also to exploit the data and GPS coordinates of their devices [26].

7.7 CONCLUSION AND FUTURE OF IIOT SYSTEMS

As the revolution of Industry 4.0 moves forward and IIoT systems continue transforming the industry for efficiency and automation, the risk of cyber-attacks and many more new challenges also grows. Implementing industry and security standards should be the new norm moving forward, which can help IIoT systems protect themselves against growing cyber threats and insider breaches. Interconnectivity among the devices and sensors that operate to enhance the industry efficiency needs to be protected, along with critical data and personal information, through adherence to various standards and control mechanisms for device operability.

Despite many challenges in maintaining security standards, constant adoption, improvements, and validation of the security standards can help organizations fight off cyber threats. This can also open the door for many new industrial possibilities such as the Product-as-a-Service (PaaS) business model, which involves selling a product as a service rather than the physical product, which can be remotely enabled to monitor and manage products, using new AI tools for predictive maintenance of systems and devices, new data analysis and data protection, encryption models which can help prevent further data breaches and find loopholes in the data collection system, which in long term can reduce cost for the organizations.

However, the adoption of IIoT systems also raises ethical considerations. For example, IIoT systems can collect vast amounts of data on employees, customers, and other stakeholders. This data must be handled carefully to protect privacy rights and prevent unauthorized access. Additionally, IIoT systems can give rise to new forms of discrimination, such as algorithmic bias, which must be addressed to ensure equality.

Future digital transformations will heavily rely on IIoT devices, particularly as businesses work to automate their supply chains and production processes. Big data analytics will also advance to include IIoT data. Organizations will be able to recognize changing conditions in real time and take appropriate action as a result [27].

REFERENCES

1. Buja, A., Apostolova, M., Luma, A., & Januzaj, Y. (2022, June). Cyber security standards for the industrial Internet of Things (IIoT) – A systematic review. In 2022 *International Congress on Human-Computer Interaction, Optimization and Robotic Applications (HORA)* (pp. 1–6). IEEE, Piscataway, NJ.
2. CISA. (2023). Shifting the Balance of Cybersecurity Risk: Principles and Approaches for Security-by-Design and – Default. Retrieved from https://www.cisa.gov/sites/default/files/2023-04/principles_approaches_for_security-by-design-default_508_0.pdf
3. TrendMicro. (2015). Industrial Internet of Things (IIoT) – Definition – Trend Micro USA. Retrieved from Trendmicro.com website: https://www.trendmicro.com/vinfo/us/security/definition/industrial-internet-of-things-iiot
4. Balbix. (2020, December 16). What is the NIST Cybersecurity Framework? Retrieved from Balbix website: https://www.balbix.com/insights/nist-cybersecurity-framework/
5. Kobes, P. (n.d.). White Paper Excerpt: Applying ISO/IEC 27001/2 and the ISA/IEC 62443 Series for Operational Technology Environments. Retrieved from gca.isa.org website: https://gca.isa.org/blog/white-paper-excerpt-applying-iso/iec-27001/2-and-the-isa/iec-62443series-for-operational-technology-environments
6. ISA. (n.d.). ISA/IEC 62443 Series of Standards – ISA. Retrieved from isa.org website: https://www.isa.org/standards-and-publications/isa-standards/isa-iec-62443-series-ofstandards
7. Synopsys. (n.d.). What Is UL 2900 and How Does It Work? | Synopsys. Retrieved from www.synopsys.com website: https://www.synopsys.com/glossary/what-is-ul-2900.html
8. Nadeau, M. (2020, June 12). General Data Protection Regulation (GDPR): What you need to know to stay compliant. Retrieved from CSO Online website: https://www.csoonline.com/article/3202771/general-data-protection-regulation-gdprrequirements-deadlines-and-facts.html
9. Uberoi, A. (2022, October 25). IoT Security: 5 Cyber-Attacks Caused by IoT Security Vulnerabilities. Retrieved from www.cm-alliance.com website: https://www.cmalliance.com/cybersecurity-blog/iot-security-5-cyber-attacks-caused-by-iot-securityvulnerabilities.
10. Cohen, G. (2021, April 22). Throwback Attack: WannaCry ransomware takes Renault-Nissan plants offline. Retrieved from Industrial Cybersecurity Pulse website: https://www.industrialcybersecuritypulse.com/facilities/throwback-attack-wannacryransomware-takes-renault-nissan-plants-offline/
11. TrendMicro. (2021, September 28). IoT and Ransomware: A Recipe for Disruption – Security News – Trend Micro PH. Retrieved April 19, 2023, from www.trendmicro.com website: https://www.trendmicro.com/vinfo/ph/security/news/internet-of-things/iot-and-ransomware-arecipe-for-disruption
12. Hassanzadeh, A., Rasekh, A., Galelli, S., Aghashahi, M., Taormina, R., Ostfeld, A., & Banks, M. K. (2020). A review of cybersecurity incidents in the water sector. *Journal of Environmental Engineering, 146*(5), 03120003.

13. Jones, C. (2021, October 26). Warnings (& Lessons) of the 2013 Target Data Breach. Retrieved from Red River | Technology Decisions Aren't Black and White. Think Red. website: https://redriver.com/security/target-data-breach #:~:text=What%20Happened%20During%20the%20Target

14. ITRC. (n.d.). Ring Doorbell Data Leak Exposes Over 3,000 Accounts. Retrieved April 19, 2023, from ITRC website: https://www.idtheftcenter.org/ post/ring-doorbell-data-leak-exposes-over3000-accounts/

15. DigitalGuardian. (n.d.). What is Data Loss Prevention (DLP)? Definition, Types & Tips. Retrieved from Digital Guardian website: https://www.digitalguardian. com/blog/what-data-lossprevention-dlp-definition-data-loss-prevention

16. Cypress Data Defense. (2020, September). How to Prevent Data Loss: 13 Simple Ways. Retrieved from www.cypressdatadefense.com website: https:// www.cypressdatadefense.com/blog/how-to-prevent-data-loss/

17. Jaytech. (2021). NIST Cybersecurity Framework – JayTech Cyber Solution. Retrieved April 20, 2023, from Jaytechcybersolution.com website: https:// www.jaytechcybersolution.com/nistcybersecurity-framework/

18. Gordon, J. (2021, December 26). The Essential Guide to the IEC 62443 industrial cybersecurity standards. Retrieved from Industrial Cyber website: https://industrialcyber.co/essentialguides/the-essential-guide-to-the-iec-62443-industrial-cybersecurity-standards/

19. CISA. (2022). Defining Insider Threats | CISA. Retrieved from www.cisa.gov website: https://www.cisa.gov/defining-insider-threats

20. IEC. (n.d.). Understanding IEC 62443. Retrieved from www.iec.ch website: https://www.iec.ch/blog/understanding-iec-62443

21. IICONSORTIUM. (n.d.). Managing Risk. Retrieved April 19, 2023, from hub.iiconsortium.org   website: https://hub.iiconsortium.org/portal/IISF_ Part2/5b6db134266d89000f8767f8

22. EATON. (n.d.). The Internet of Things and the Energy Sector: Myth or Opportunity. Retrieved from EATON website: https://www.eaton.com/ us/en-us/markets/utilities/knowledge-center/IoTand-energy-myth-or-opportunity.html

23. Ryan, M. (2022). *Ransomware Revolution: The Rise of a Prodigious Cyber Threat* (1st ed., Vol. XV, 156, pp. XV, 156). Springer, Cham. Retrieved from https://doi.org/10.1007/978-3-030-66583-8 (Original work published 2021).

24. CyberTalk.org. (n.d.). Prevention and Security Architecture Keep Transportation on the Right Track. Retrieved from CyberTalk.org website: https://www.cyber talk.org/prevention-andsecurity-architecture-keep-transportation-on-the-right-track/

25. Morrow, S. (2022, March 7). What are top Cybersecurity threats in the Manufacturing sector? Retrieved April 19, 2023, from www.netify.com website: https://www.netify.com/learning/whatare-cybersecurity-threats-in-the-manufacturing-sector

26. Jeremy, K. (2021, August 9). Flaws in John Deere Systems Show Agriculture's Cyber Risk. Retrieved from www.bankinfosecurity.com website: https:// www.bankinfosecurity.com/flaws-in-john-deere-systems-show-agricultures-cyber-riska-17240

27. RYUKK. (2022, June 10). Industrial Internet of Things – IIoT. Retrieved April 18, 2023, from medium website: https://medium.com/@mohammad. raza20/industrial-internet-of-things-iiot5363c0be1146

Chapter 8

Intrusion detection system and its types in IoT-enabled devices

Ilemona Atawodi and Zhaoxian Zhou

8.1 INTRODUCTION

Fundamentally, the Internet of Things (IoT) refers to a network of interconnected devices that communicate with each other on a common network, generating and exchanging data with a high degree of autonomy. These devices could range from smart devices that interact directly with human beings, like smartwatches and fitness trackers, to home-centered devices like smart speakers, internet-enabled thermostats, and microwaves. They could also be part of public infrastructure, enabling smart city technology. IoT devices can be found in every aspect of life, and the number of ways IoT devices are being utilized increases every day.

The popular open-source hardware platforms known as Arduino and Raspberry Pi have also grown to empower a lot of individuals to come up with very creative IoT devices that address very particular problems unique to these individuals. These solutions are often shared with a very active community of hobbyists, leading to the adoption of these unique IoT devices. A good example of this is a smart collar proposed by the author of this project [1], which could be used to wirelessly track livestock on a farm in real time, making use of simple microcontrollers, some sensors, and a small transmitter. A major reason behind the fast growth and adoption of IoT devices is their flexibility. There are ready-made solutions for certain use cases, such as smart home cameras and speakers like the Google Nest solutions, but individuals also have the freedom to build out solutions that suit their particular unique use cases in ways tailored to their own needs.

Considering that this has enabled IoT devices to become a big part of every facet of human life, this makes IoT-enabled devices a prime target for cybercriminals. For instance, an IoT-enabled indoor camera could be hijacked, invading the privacy of the user to blackmail them or a delivery truck making use of an IoT tracking device could have its IoT-enabled devices hijacked for nefarious purposes, and other malicious actors could attack critical infrastructure to inflict financial losses on a competitor or even compromise a nation's security. This highlights the need to ensure that

DOI: 10.1201/9781003530572-8

these devices operate securely, safeguarding our safety and privacy as we go about our daily lives. An effective approach to securing networked devices is to implement intrusion detection systems (IDS).

8.1.1 Importance of intrusion detection systems

The first and most important reason for employing an intrusion detection system in a network is the ability of IDS to rapidly identify and respond to security incidents. IDS scan and monitor network traffic and device activity for anomalous behavior, alerting cybersecurity administrators to potential threats in real time, and enabling quick responses to potential security incidents. Both homes and businesses face an ongoing risk of cyberattacks, and having the means to raise alarms in the event of a potential attack goes a long way in mitigating any potential issues before they escalate to catastrophic levels.

The second reason would be that IDS help both homes and organizations improve their security posture. This means that utilizing an IDS can aid in identifying vulnerabilities in your network and devices, which could guide the individual or organization to implement mitigation measures. Implementing these mitigation methods in turn improves the overall security presence of the network and makes it less vulnerable to attacks.

These reasons are general and can be applied to any network security situation. There are benefits that IDSs present that are particularly beneficial to IoT networks considering the fact that their architecture makes them more vulnerable to cyberattacks than regular networks.

The first reason is that IoT-specific vulnerabilities exist and are often exploited by attackers. IDS can be tailored to the specific environment in which they are deployed. Therefore, they can be configured to detect certain attacks that are prevalent in IoT devices and networks. This is a significant advantage because a general-purpose detection tool might protect against common situations but could miss many IoT-specific incidents.

Secondly, IoT devices are often tasked with handling, reading, and transmitting sensitive data such as medical, financial, or personally identifiable information such as tracking data and geolocation. This makes them a prime target for large-scale hacks that often lead to data breaches. IDSs have long been a useful tool in protecting users from becoming victims of such breaches.

Overall, it is important to note that IoT devices generate a large volume of network traffic, making it difficult for security teams to monitor the networks for suspicious activity involving IoT devices. Tools like IDSs help security teams to maintain a high level of network security when IoT devices are deployed, as they are more vulnerable to attacks compared to other networks without them.

8.2 TYPES OF SECURITY THREATS

There are numerous options for networking IoT devices, along with a wide range of building blocks and deployment strategies. This diversity in deployment and connections also introduces a variety of attack types that can affect an IoT setup. In this subsection, we will cover three primary classes of IoT attacks (Figure 8.1):

- Data Access/Manipulation
- Disruption/Malware
- Physical Attacks

8.2.1 Data access/manipulation

The class of Data Access/Manipulation attacks encompasses cyberattacks aimed at compromising data stored within computer systems and networks. These attacks may involve unauthorized actions such as data theft, alteration, deletion, or the obstruction of authorized users' access to data. This attack class comprises of:

- *Data Breaches*: Data breaches involve unauthorized access to sensitive information and can manifest through various means, including hacking, malware infections, and human error. This attack type is often attributed to a lack of cybersecurity awareness and education. In many cases, organizations invest significantly in educating their staff

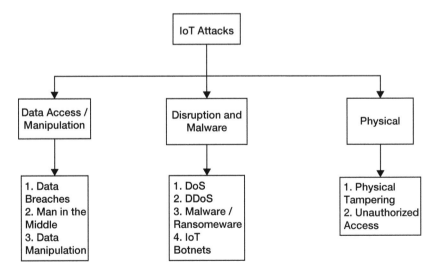

Figure 8.1 Types of attacks.

about actions and habits that could expose the organization to data breaches. This is especially critical since the primary route followed by cybercriminals for this kind of attack is often social engineering.

- *Man-in-the-Middle (MitM) Attacks*: MitM attacks revolve around the interception and manipulation of communications between two parties. They serve as a means to either steal data, alter data, or impersonate one of the communicating parties. This kind of attack is very common in public places that offer free WiFi connections but is not limited to WiFi connections, any kind of connection between two or more parties can be intercepted with a MitM attack. For example, the connection between a sensor node and a command node could be intercepted, and the attackers may use that hijacked line to enact their goals.

- *Data Manipulation Attacks*: Data manipulation attacks involve unauthorized alterations to data, and they can take various forms. These attacks may be employed to corrupt data, modify data to conceal malicious activity, or create false data with the intent to deceive or disrupt. This could also be done for financial gain.

8.2.2 Disruption and malware attacks

Disruption and Malware Attacks are a class of cyberattacks that aim to disrupt the operation of computer systems and networks or to install malicious software on those systems. These attacks can cause a variety of problems, including outages, data loss, and performance degradation. Attacks under this class comprise:

- *Denial-of-Service (DoS) Attacks*: DoS attacks are engineered to flood a system with excessive traffic, rendering it inaccessible to legitimate users. These attacks can be directed at databases, servers, and various network devices, including IoT devices. Many homes use smart locks these days. If a home is hit with a DoS attack, it could deny the members of that home access due to the lock being unavailable to the owner, even though they have legitimate access.

- *Distributed Denial-of-Service (DDoS) Attacks*: DDoS attacks are similar to DoS attacks, but they involve using a large number of compromised devices to generate the traffic. DDoS attacks are often much more powerful than DoS attacks, and they can be difficult to defend against due to the sheer amount of devices involved in this kind of attack.

- *Malware and Ransomware Attacks*: Malware attacks involve installing malicious software on computer systems and networks. Malware can be used to steal data, corrupt data, or disrupt the operation of systems. On the other hand, ransomware is used to disrupt the operations of a network, steal data, or withhold access from the legitimate administrators of a network to demand a ransom in exchange for access back into their network or database.

- *IoT Botnets*: An IoT botnet is a network of IoT devices that have been infected with malware and have fallen victim to a malicious actor. IoT botnets can be used to launch a variety of attacks, including DDoS attacks and data breaches. IoT botnets are particularly dangerous because IoT devices are often insecure and have weak passwords. Additionally, IoT devices are often always connected to the internet, which gives attackers a large army of potential targets to choose from. IoT botnets can also be used to utilize the hijacked IoT devices to mine cryptocurrencies for the attacker without the owner's consent. IoT botnets can be difficult to detect and remove because they are often distributed across a large number of devices.

8.2.3 Physical attacks

Physical attacks are cyberattacks that involve physical access to computer systems and networks. These attacks can be used to steal data or hardware, damage hardware, or create a financial burden on the target due to the need to replace lost equipment. This class of attack is often simple but can still have potentially problematic end results. For example, IoT devices like smart cameras may have embedded or removable memory cards containing private or personal data which could have grave consequences for the entity under attack. These cards could be exchanged with a clone of the device, essentially acting as a trojan horse on the network. This could later be used to eavesdrop, tamper with the network, or gain further access to deeper points in the network.

8.3 TYPES OF INTRUSION DETECTION SYSTEM

There are five general types of IDS, which can further be categorized into two classes based on data source and detection method (see Section 8.2, Figure 8.2).

Before discussing the types of IDS, it is important to understand the architecture of an intrusion detection system. An IDS is made up of basically five parts [2,3].

1. *IDS Sensor/Agent*: An IDS sensor or agent is a part of the IDS that is installed on a network or host to oversee its operations. Sensors monitor both wired and wireless networks, while agents focus on tracking endpoints or hosts. The quantity of sensors or agents you need to deploy depends on the size of your network or the number of hosts

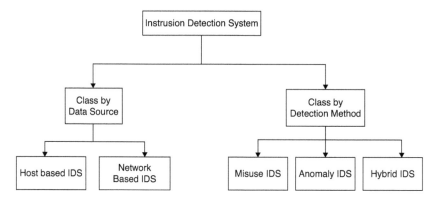

Figure 8.2 Types of intrusion detection systems.

you wish to monitor. Sensors can be deployed on network devices such as routers and switches, or individual devices like laptops and servers.

2. *Management Server/Collector*: An IDS management server assumes the task of gathering data from both sensors and agents. It also holds the capacity to correlate and analyze this collected data. In some IDS solutions, you may come across setups with one or more management servers, whereas others operate without any. Extensive deployments, such as those in large enterprise networks, frequently employ multiple management servers, while smaller deployments, as seen in small businesses, typically forgo the use of management servers.

3. *Database*: It is important for an IDS to have a database as a part of its architecture, providing a place to store relevant event data and logs from the IDS sensors and agents.

4. *Alerting System*: The alerting system analyzes the collected data to detect any unusual or suspicious activities. This system employs a range of techniques for detecting suspicious behavior, including signature-based detection, anomaly-based detection, and machine learning-based detection. Once the collector detects any such suspicious activity, the alerting system promptly notifies the security administrator.

5. *IDS Console*: The IDS console serves as the hub for carrying out administrative or management functions. How you interact with the IDS console depends on the vendor's specific design. It may be used for tasks such as configuring sensors or agents, as well as for monitoring and analysis. Some IDS consoles offer capabilities for both configuring sensors or agents and performing monitoring and analysis, while others are limited to configuring sensors or agents only.

8.3.1 IDS classified by detection method

8.3.1.1 Signature-based detection

Signature-based IDS compare network traffic or device activity against a database of known attack signatures. They are effective at identifying known attacks but are unable to detect new or previously unknown attacks. Signature-based detection operates by comparing network traffic or device activity to a database of attack signatures, typically created by security researchers who analyze known attacks and develop patterns for future identification.

When a signature-based IDS identifies a match with a known attack signature, it triggers an alert. A security administrator can then use this alert to investigate the potential attack and take appropriate action. Signature-based detection is a relatively straightforward and easily implemented IDS technology, particularly effective in identifying known attacks [4].

8.3.1.2 Anomaly-based detection

Anomaly-based IDS examine network traffic or device activity to identify deviations from normal behavior. They are effective at detecting new and unknown attacks but can sometimes generate false positives. Anomaly-based detection functions by establishing a baseline of normal behavior for network traffic or device activity. This baseline is created by gathering data over a period and identifying patterns within it. Once the baseline is established, the anomaly-based IDS can monitor network traffic or device activity for deviations from this norm [4].

When an anomaly-based IDS detects a deviation from the baseline, it triggers an alert. Security personnel can then use this alert to investigate the potential attack and take appropriate action. Anomaly-based detection is a more complex IDS technology compared to signature-based detection. It is also more challenging to implement and manage. However, anomaly-based detection offers certain advantages over signature-based detection, especially since it can detect zero-day attacks and is significantly more difficult for attackers to bypass [4,5].

8.3.1.3 Machine learning-based detection

A machine learning-based intrusion detection system leverages machine learning algorithms to train models capable of accurately identifying malicious activity. It is effective against both known and unknown attacks, although it may require more resources and training data than other detection methods [6,7]. Machine learning-based detection operates by training a machine learning model using a dataset comprising known attacks and benign activity. The model learns to recognize patterns in the data associated with malicious behavior. Once trained, the model can scan network traffic or monitor device activity for these distinctive patterns. When a machine learning-based IDS identifies a pattern linked to malicious activity, it alerts security personnel to investigate the potential attack and take appropriate measures [4,6].

Machine learning-based detection represents a more sophisticated IDS technology than signature-based or anomaly-based detection. It is also more resource-intensive and necessitates a larger volume of training data. Nevertheless, machine learning-based detection offers several advantages over other IDS technologies as it effectively combats both known and unknown attacks.

8.3.2 IDS classified by data source

When IDSs are classified by data source, you end up with two types: host-based IDS and network-based IDS. It is important to note that both of these classes function based on their data source and employ one of the approaches mentioned above.

8.3.2.1 Host-based intrusion detection

A host-based intrusion detection system (HIDS) is a type of IDS that monitors a single computer system for suspicious activity. HIDS software runs on the host system and continually monitors its logs, system files, and activities for signs of suspicious behavior or security threats. This can include unauthorized access, file system changes, and abnormal processes.

HIDS is a critical component of a comprehensive network security strategy because it focuses on the security of individual computers, servers, or devices. It is installed directly on the host it is meant to protect. In addition to detecting security incidents, some HIDS solutions are designed to respond to threats by taking actions such as blocking suspicious network traffic or isolating the host from the network.

It can be customized to focus on the specific security needs of the host it is protecting, allowing for tailored security policies and alerts. HIDS can employ various methods to detect anomalies or deviations from established baseline behavior. This may include signature-based detection, which looks for known patterns of attacks, and anomaly-based detection, which identifies deviations from the host's typical behavior. HIDS is commonly used on critical servers and workstations, especially in enterprise environments, to ensure the security and integrity of individual hosts and can detect both internal and external threats.

When HIDS detects suspicious or malicious activity, it raises an alert, which can be logged and reported to the system administrator or a centralized security console, as mentioned in the IDS architecture section.

8.3.2.2 Network-based intrusion detection

A Network Intrusion Detection System (NIDS) is a security solution designed to monitor and analyze network traffic for signs of malicious activity or security breaches. It continuously inspects network traffic, looking for patterns or anomalies that may indicate a security threat. Operating at the network level, it analyzes data packets as they traverse the network, including data packets, headers, and protocols. NIDS can monitor traffic within a local area network (LAN), wide area network (WAN), or on the internet [8].

In a signature-based detection scheme, the NIDS compares network traffic to a database of known attack patterns or signatures. If it identifies a match, it generates an alert. An anomaly-based detection scheme looks for deviations from established baselines of network behavior. It identifies abnormal patterns, which may suggest an attack. When a NIDS identifies a potentially malicious activity, it generates alerts that can be sent to a central management console for analysis. In some cases, the NIDS can be configured to take predefined response actions, such as blocking traffic from a suspicious IP address or isolating a compromised segment of the network. NIDS can be deployed in two primary modes:

- *Passive Mode*: In passive mode, a NIDS monitors network traffic without interfering with it. It acts like an observer, providing alerts but not taking active control.
- *Inline Mode*: In inline mode, a NIDS is placed directly in the network traffic path. It can actively block or drop suspicious traffic in real time.

Network IDSs can be placed at strategic points in the network, such as at network gateways or within specific network segments, to provide visibility into different parts of the network. They play a vital role in detecting a wide range of network-based threats, including intrusion attempts, malware propagation, DoS attacks, and unauthorized access to network resources.

Network IDSs are often used in conjunction with other security measures such as firewalls, intrusion prevention systems (IPSs), and HIDSs to provide comprehensive network security. They are suitable for organizations of all sizes and industries and are widely used to protect networks from cyber threats and security breaches [9].

8.4 DEPLOYMENT CONSTRAINTS

One major challenge that tends to plague IoT-enabled devices usually stems from their low-power design. IoT devices tend to operate with very little computational resources and memory. This is usually an advantage in terms of power consumption since they tend to be always online, transmitting and receiving data, but it restricts the devices and the kinds of additions that can improve the security of these devices. Another issue is that since most IoT software is burned onto the IoT SoC, they are generally always running old and vulnerable software. The aforementioned constraints are explained further below:

- *Resource Constraints*: IoT devices typically have limited processing power, memory, and storage resources. This can make it difficult to deploy traditional IDSs on IoT devices, which are often designed to be as lightweight and efficient as possible.
- *Power Consumption*: IoT devices are often powered by batteries or low-power energy sources. This means that IDSs deployed on IoT devices must be designed to minimize power consumption to avoid draining the battery too quickly.
- *Complexity*: IDSs can be complex to configure and manage. This complexity can make it difficult to deploy IDSs on IoT devices, which are often managed by non-security experts.

8.5 FUTURE TRENDS FOR IDS IN IOT-ENABLED DEVICES

With chips getting smaller and smarter [10] and capable of doing a lot more with very little, the future of IDS in IoT would bring a lot of innovations to the technology. Some of these possible technologies are increased use of artificial intelligence; AI-based IDSs are already being used to detect a wide range of threats in IoT devices, including malware, DoS attacks, and unauthorized access. A good example is the company Darktrace [11] which developed an AI-based IDS that can detect malware in IoT devices by analyzing their network traffic and behavior patterns. The use of blockchain technology [12] has the possibility to improve the resilience of IoT against tampering and enable better integration with existing security solutions. Blockchain can also facilitate the sharing of threat intelligence between IDSs, thereby improving the overall security of IoT networks [13]. Finally, edge computing can deploy IDSs closer to the IoT devices they protect [14]. This can reduce latency and improve performance, especially for critical IoT applications.

REFERENCES

1. B. Risteska Stojkoska, D. Capeska-Bogatinoska, G. Scheepers, and R. Malekian, "Real-time Internet of Things architecture for wireless livestock tracking," *Telfor Journal*, vol. 10, no. 2, pp. 74–79, 2018.
2. B. P. Brien, "Ids: A comprehensive guide to intrusion detection systems." https://techgenix.com/ids-intrusion-detection-system-guide/, 2023. Accessed: 2023-10-15.

3. M. Garuba, C. Liu, and D. Fraites, "Intrusion techniques: Comparative study of network intrusion detection systems," in *Fifth International Conference on Information Technology: New Generations (ITNG 2008)*, pp. 592–598, IEEE, Las Vegas, NV, 2008.

4. I. S. Atawodi, "A machine learning approach to network intrusion detection system using k nearest neighbor and random forest," Masters Thesis, University of Southern Mississippi, 2019.

5. V. Jyothsna and K. M. Prasad, "Anomaly-based intrusion detection system," *Computer and Network Security*, vol. 10, p. 35, 2019.

6. Q. Niyaz, W. Sun, A. Y. Javaid, and M. Alam, "A deep learning approach for network intrusion detection system," in *Proceedings of the 9th EAI International Conference on Bio-inspired Information and Communications Technologies (Formerly BIONETICS), BICT-15*, New York City, New York, United States, vol. 15, pp. 21–26, 2015.

7. M. Sarhan, S. Layeghy, and M. Portmann, "Towards a standard feature set for network intrusion detection system datasets," *Mobile Networks and Applications*, vol. 27, pp. 1–14, 2022.

8. Z. Ahmad, A. Shahid Khan, C. Wai Shiang, J. Abdullah, and F. Ahmad, "Network intrusion detection system: A systematic study of machine learning and deep learning approaches," *Transactions on Emerging Telecommunications Technologies*, vol. 32, no. 1, p. e4150, 2021.

9. UpGuard, "IDS vs IPS: What's the difference?," 2023. Accessed: 2023-04-04.

10. W. Knight, "Risc-v: The open-source computer chip that could revolutionize the industry." https://www.technologyreview.com/2023/01/09/1064876/riscvcomputer-chips-10-breakthough-technologies-2023/, 2023. Accessed: 2023-10-15.

11. F. Piconese, A. Hakkala, S. Virtanen, and B. Crispo, "Deployment of next generation intrusion detection systems against internal threats in a medium-sized enterprise," Masters Thesis, University of Turku, Finland, 2020.

12. A. Dorri, S. S. Kanhere, R. Jurdak, and P. Gauravaram, "LSB: A lightweight scalable blockchain for iot security and anonymity," *Journal of Parallel and Distributed Computing*, vol. 134, pp. 180–197, 2019.

13. A. I. Sanka and R. C. Cheung, "A systematic review of blockchain scalability: Issues, solutions, analysis and future research," *Journal of Network and Computer Applications*, vol. 195, p. 103232, 2021.

14. Y. Wang, W. Meng, W. Li, Z. Liu, Y. Liu, and H. Xue, "Adaptive machine learningbased alarm reduction via edge computing for distributed intrusion detection systems," *Concurrency and Computation: Practice and Experience*, vol. 31, no. 19, p. e5101, 2019.

Chapter 9

IIoT edge network security

Addressing spectrum scarcity through intrusion detection systems – a critical review

Md. Alimul Haque, Deepa Sonal,
Shipra Shivkumar Yadav, Shameemul Haque,
Anil Kumar Sinha, and Sultan Ahmad

9.1 INTRODUCTION

The Internet of Things (IoT) has become an increasingly popular target for cybercriminals due to the sheer number of devices connected to the internet and the lack of security in many of these devices. One of the main reasons why IoT devices are vulnerable to cyber-attacks is that they often lack the necessary security measures to protect against unauthorized access (Granjal et al., 2015; Haque, Haque, Kumar, et al., 2023). For example, many devices come with default login credentials that users do not change, making it easy for attackers to gain access. Hackers can exploit these vulnerabilities to gain access to the device or its data, launch Distributed Denial-of-Service (DDoS) attacks, or even use the device as a launchpad for further attacks on other devices or networks. Another issue is the lack of encryption in communication between IoT devices and their associated apps or cloud services. This means that data transmitted between devices may be intercepted and stolen by cybercriminals, compromising user privacy and security. As IoT devices become more prevalent in our lives, manufacturers and users need to prioritize security and take steps to protect against cyber-attacks. To protect IoT devices from these attacks, it is important to take steps such as changing default login credentials, keeping devices up to date with the latest security patches, and using strong encryption methods to protect sensitive data. This includes regularly updating software and firmware, changing default login credentials, implementing encryption, and following best practices for network security.

9.1.1 Internet of Things (IoT)

The term "Internet of Things" (IoT) denotes a collection of physical objects or "things" that incorporate sensors, software, and additional technologies,

DOI: 10.1201/9781003530572-9

facilitating their ability to interconnect and exchange data with other systems and devices over the internet. These objects encompass a broad range, from intelligent household appliances and wearables to industrial machinery and transportation networks. The IoT is driven by the growing number of connected devices and the increasing availability of high-speed internet connectivity (Haque, Haque, Zeba, et al., 2023; Haque, Almrezeq, et al., 2022). By connecting these devices to the internet, they can be remotely monitored, managed, and controlled, creating new opportunities for automation and optimization.

Some of the key benefits of IoT include:

- *Increased Efficiency:* IoT devices can help organizations optimize their operations and reduce costs by automating tasks and improving resource utilization.
- *Improved Decision-Making:* IoT devices can assist organizations in making more informed and timely decisions by providing real-time data and analytics.
- *Enhanced Customer Experience:* IoT devices can be used to create personalized and interactive experiences for customers, improving engagement and satisfaction.
- *Improved Safety and Security:* IoT devices can be used to monitor and control physical environments, improving safety and security in various settings.

However, the IoT also poses significant cybersecurity challenges. With so many connected devices, there are more potential entry points for cyber-attacks, and many IoT devices have weak security protocols and are vulnerable to exploitation. Ensuring the security and privacy of IoT devices is critical to realizing the full potential of the IoT while minimizing the risk of cyber threats.

The main objective of this chapter is to provide a comprehensive review of intrusion detection systems (IDS) in the context of IoT. The chapter covers various aspects of IDS in IoT, including different techniques used for intrusion detection, deployment strategies, types of attacks on IoT networks, and public datasets available for evaluating IDS.

Overall, the chapter provides a valuable resource for researchers, practitioners, and policymakers interested in understanding the state of the art in IDS for IoT and the challenges associated with securing IoT networks (Khraisat et al., 2019; Hindy et al., 2018; Benkhelifa et al., 2018; Chaabouni et al., 2019; Zarpelão et al., 2017). The review is based on an extensive analysis of existing literature in the field, and the authors have made significant contributions to the existing knowledge by providing a critical analysis of current research trends and gaps in the literature. The authors aim to identify the strengths and weaknesses of existing IDS techniques for IoT and highlight the major challenges faced by researchers in this area.

This chapter focuses on the following:

- Categorizing IoT IDS into different groups based on intrusion methods.
- Presenting recent efforts to enhance the security of IoT IDS.
- Introducing a classification system for IoT attacks.
- Providing an overview of existing IDS datasets.
- Discussing difficulties associated with implementing IoT IDS.

9.2 LITERATURE SURVEY

A literature survey of the IoT can provide an overview of the current state of research, key trends, challenges, and potential applications of the technology. Here are some key themes and findings from recent IoT literature:

Abbasi and J. Wetzels propose an emulation-based approach to network intrusion detection. The idea is to create a virtual environment that mimics the behavior of the target system and then run potentially malicious network traffic through the environment to observe its behavior. This approach can detect attacks that have not been previously seen and can accurately distinguish between malicious and benign traffic.

The authors present the design and implementation of their emulation-based IDS, which consists of three main components: the emulation engine, the network capture component, and the analysis component. The emulation engine is responsible for creating a virtual environment that mimics the target system, while the network capture component intercepts network traffic and feeds it into the emulation engine. The analysis component then examines the behavior of the traffic in the virtual environment to determine whether it is malicious or benign.

To evaluate their approach, the authors conducted experiments using both real-world and synthetic datasets. The results showed that their emulation-based IDS achieved higher accuracy and lower false positive rates compared to traditional IDS. The authors also discuss the limitations of their approach and potential future directions for research in this area (Abbasi et al., 2014).

The authors begin by discussing the limitations of traditional IDS that use a single classifier, such as high false positive rates and the inability to detect previously unseen attacks. The paper provides an overview of different ensemble and hybrid classifiers, including bagging, boosting, stacking, and fusion-based approaches. The authors then review several IDS that use ensemble and hybrid classifiers, including those specifically designed for wireless sensor networks, cloud computing environments, and industrial control systems. They also discuss the challenges and future directions for research in this area, such as developing more efficient and effective ensemble and hybrid classifiers and integrating them with other security mechanisms (Aburomman & Reaz, 2017).

The article begins by defining what is meant by an anomaly, which is an observation that deviates significantly from what is expected or normal. The authors then discuss the importance of anomaly detection in various fields such as finance, security, and healthcare. Next, the article provides a survey of various data mining techniques used for anomaly detection, including statistical methods, clustering, classification, association rules, and outlier analysis. For each technique, the authors explain the basic principles and provide examples of how the technique has been used in practice (Agrawal & Agrawal, 2015).

The authors begin by providing background information on malware and the challenges associated with detecting and classifying it. They then introduce HMMs and explain how these models can be used to characterize malware behavior based on the sequence of system calls made by the malware. Next, the authors describe their experimental setup and evaluation methodology, which involves using a dataset of real-world malware and benign software to train and test the HMM-based classifier. The article provides a valuable contribution to malware detection and classification by demonstrating the effectiveness of HMMs as a tool for this purpose. It is a useful resource for researchers and practitioners interested in applying machine learning techniques to cybersecurity (Annachhatre et al., 2015).

The authors collected the data by creating a test bed consisting of various IoT devices and infecting them with known botnet malware. They then captured the network traffic generated by the infected devices and labeled it according to the type of malware used. The resulting dataset includes traffic from multiple botnets, including Mirai, Gafgyt, and Bashlite (Koroniotis et al., 2019).

The authors also demonstrate the utility of the dataset by using it to train and evaluate a machine learning model for botnet detection. The Bot-IoT dataset is a valuable resource for researchers and practitioners in the field of network security and provides a realistic and diverse dataset for the development and evaluation of botnet detection techniques in IoT environments (Bzai et al., 2022).

The paper by Kreibich and Crowcroft (2004) presents Honeycomb, a system for creating intrusion detection signatures using honeypots. The authors propose the use of honeypots, which are systems designed to be attractive targets for attackers, to capture and analyze the behavior of attackers in order to develop new intrusion detection signatures.

The Honeycomb system consists of a distributed network of honeypots designed to emulate different types of systems and services. The honeypots are configured to log all activity, including both successful and unsuccessful attacks, and the resulting data is analyzed to identify behavior patterns. The authors demonstrate the effectiveness of Honeycomb by using it to develop signatures for detecting new and unknown types of attacks. They show that the system can identify attacks that existing signature-based systems cannot detect and that the resulting signatures have a low false positive rate. The Honeycomb system, including its architecture, data collection and analysis techniques, and signature generation process, is discussed in this paper. The authors also address the limitations of the system, such as the need for a large number of honeypots and the potential for attackers to detect and avoid honeypots. In conclusion, the paper presents a novel approach to intrusion detection using honeypots to capture and analyze the behavior of attackers. The Honeycomb system effectively generates new intrusion detection signatures and has the potential to enhance the accuracy of existing signature-based systems (Kreibich & Crowcroft, 2004).

The integration of IoT devices in education has enabled the development of e-learning systems that provide personalized learning experiences to students. However, these systems face challenges related to security, privacy, and data ownership. Blockchain technology offers a potential solution to these challenges by providing a decentralized and secure platform for data sharing and management. This would enable the creation of a shared database of educational records that is tamper-proof and accessible only by authorized parties. Blockchain technology can also help ensure the authenticity of educational certificates and degrees, which can be verified using a public blockchain ledger. Moreover, the use of blockchain technology can help reduce the administrative costs associated with traditional educational systems by streamlining processes such as enrollment, grading, and certification. This would enable educational institutions to focus on providing high-quality education to students while reducing overhead costs. The integration of blockchain technology in E-learning IoT systems can help improve the efficiency and sustainability of educational systems while providing secure and transparent data-sharing capabilities (Haque, Sonal, et al., 2022).

The proposed system uses machine learning algorithms to analyze student data, such as their learning history and performance, to generate personalized recommendations for learning materials and activities. The system also includes features for tracking student progress and providing real-time feedback to both students and teachers.

The authors demonstrate the effectiveness of the system by evaluating its performance on a dataset of student records from a university. The results show that the proposed system can provide personalized recommendations that are more accurate than those of traditional LMSs and that the system can help improve student engagement and performance (Haque et al., 2021).

The proposed system consists of a network of IoT devices, including cameras and sensors, that are deployed in examination halls. The devices are designed to capture and analyze real-time data such as images and sounds to detect any suspicious activity. The data is then analyzed using machine learning algorithms to identify potential malpractices. The authors demonstrate the effectiveness of the proposed system by evaluating its performance on a dataset of simulated exam scenarios. The results show that the system can accurately detect potential malpractices with a high degree of accuracy and can improve the security and integrity of examinations (Haque, 2021).

The paper by Haque (2021a) discusses how blockchain technology can provide a secure and decentralized platform for data sharing and management, which can help enhance the security and privacy of IoT devices.

The author also discusses various attack vectors that can be exploited in an IoT system, such as network protocols, firmware, and hardware vulnerabilities. The chapter then further discusses various approaches that can be used to classify cybersecurity attacks. The author identifies three primary classification approaches: based on the attacker's motive, based on the attack methodology, and based on the type of target system (Masud et al., 2020). The author also provides examples of each classification approach to illustrate how they can be applied in practice. Finally, the chapter discusses various countermeasures that can be implemented to mitigate the risk of cybersecurity attacks in an IoT system. These countermeasures include security measures at the device level (e.g., authentication and access control), at the network level (e.g., firewalls and IDS), and at the application level (e.g., encryption and data validation). The author also discusses the importance of keeping software and firmware up to date and implementing secure coding practices to prevent vulnerabilities from being introduced in the first place (Haque et al., 2021).

Overall, this chapter provides a comprehensive overview of the cybersecurity challenges associated with IoT and the countermeasures that can be implemented to mitigate those challenges. It serves as a valuable resource for researchers, practitioners, and policymakers working in cybersecurity and IoT.

9.3 INTRUSION DETECTION IN THE INTERNET OF THINGS

Intrusion detection in IoT is an important area of research, as the widespread adoption of IoT devices has led to new security challenges and vulnerabilities. Additionally, the large number of connected devices in the IoT creates a larger attack surface for attackers to exploit. Intrusion detection in IoT can be challenging due to the unique characteristics of IoT networks. Traditional IDS may not effectively detect IoT-specific attacks such as physical attacks, replay attacks, and jamming attacks (Srivastava et al., 2022; Ahmad, Jha, Alam, et al., 2022). These attacks can exploit the physical characteristics of IoT devices, such as their location and proximity to other devices. To address these challenges, researchers are developing new intrusion detection techniques and frameworks specifically tailored for IoT networks (Ahmad, Jha, Abdeljaber, et al., 2022). Several methods are available for detecting anomalies in IoT networks, including machine learning algorithms, network flow analysis, and behavior-based intrusion detection. Furthermore, scientists are investigating how blockchain technology could improve the security and privacy of these networks, by creating an immutable and distributed ledger of IoT transactions, ensuring it is tamper-proof and decentralized (Santos et al., 2018).

One of the key challenges in intrusion detection in IoT is the lack of standardized datasets and benchmarks for testing and evaluating intrusion detection algorithms. This can make it difficult to compare the performance of different intrusion detection techniques and to develop effective and accurate IDS for IoT networks. Intrusion detection in IoT is critical because the security of IoT networks is essential for ensuring the safety and privacy of users and protecting against cyber-attacks. As the use of IoT devices continues to grow, it becomes important to develop effective and accurate IDS that can countermeasure attacks and vulnerabilities.

9.4 IOT INTRUSION DETECTION SYSTEMS METHODS

IDS are essential components of cybersecurity in an IoT environment. Various methods are used for IoT intrusion detection, including signature-based detection, anomaly detection, and hybrid detection. Signature-based detection is used to detect potential threats by cross-referencing network traffic with a database of pre-existing attack signatures. This method is effective in identifying known attacks and is frequently employed in conventional IDS systems. However, it can be less effective in detecting new or unknown attacks (Ahmad & Afzal, n.d.).

Using machine learning algorithms, anomaly detection analyzes network traffic to uncover atypical or anomalous behavior that could indicate a security breach. This method is proficient at detecting novel or unrecognized attacks and has the ability to adjust to network changes over time. However, it can generate a higher number of false positives, as it may flag legitimate activity as anomalous (Srivastava et al., 2022).

Hybrid detection combines both signature-based and anomaly-based detection to improve the accuracy and effectiveness of intrusion detection. This approach leverages the strengths of both methods to detect a wider range of attacks while reducing the number of false positives. Other methods used for IoT intrusion detection include flow-based detection, which focuses on analyzing network flows to identify behavior patterns, and behavior-based detection, which analyzes the behavior of IoT devices and users to detect unusual or suspicious activity. In addition to these methods, researchers are exploring the use of artificial intelligence (AI) and machine learning algorithms for IoT intrusion detection. AI-based approaches can improve the accuracy and speed of intrusion detection and can adapt to new and emerging threats in real time. Overall, the choice of intrusion detection method depends on the specific requirements and characteristics of the IoT environment, including the type of devices and applications used, the network topology, and the security policies in place. Combining different methods and techniques may be necessary to achieve effective and accurate intrusion detection in the IoT environment (Qinxia et al., 2021).

A. Signature-based intrusion detection systems (SIDS)

SIDS is a commonly used method for detecting and preventing cyber-attacks. This approach involves comparing network traffic against a database of known attack signatures or patterns to identify and block malicious activity. SIDS work by scanning incoming network traffic for patterns that match known attack signatures. These signatures are predefined rules that describe the characteristics of a particular type of attack. When network traffic matches a signature, the SIDS can take action to block the traffic or alert network administrators. SIDS are effective in detecting known attacks, as the signatures are based on previously observed attacks and can be updated as new threats emerge. They are also relatively simple to implement and do not require extensive processing power or resources.

Despite its usefulness, SIDS may not be as effective at identifying new or unrecognized attacks that don't correspond with pre-established signatures. Attackers may also be able to avoid detection by altering their attack signatures or by utilizing tactics like encryption to camouflage their network traffic. To address these limitations, other intrusion detection methods, such as anomaly detection and behavior-based detection, have been developed. These methods use machine learning algorithms to analyze network traffic and identify unusual or suspicious behavior that may indicate an attack. They can detect new or unknown attacks and adapt to changes in the network over time (Symantec, 2017).

Overall, SIDS remains an important component of intrusion detection and prevention in cybersecurity. However, it should be used in conjunction with other intrusion detection methods to provide comprehensive and effective protection against cyber threats.

B. Anomaly-based intrusion detection system (AIDS)

AIDS is a type of intrusion detection system that uses machine learning algorithms to identify abnormal or anomalous behavior in network traffic. Unlike SIDS, which relies on predefined rules and signatures to detect known attacks, AIDS is designed to identify new and previously unknown types of attacks.

AIDS works by building a baseline of normal network traffic behavior over time and then comparing incoming traffic against this baseline to detect anomalies. The system may flag traffic as anomalous if it exhibits unusual patterns or behavior, such as unusual traffic volume, sources, or patterns of data transfer.

One advantage of AIDS is that they can detect new or previously unknown types of attacks that may not be covered by predefined signatures used by SIDS. They can also adapt to changes in the network over time, such as changes in traffic patterns or the introduction of new devices.

However, AIDS can generate a higher number of false positives than SIDS, as they may flag legitimate traffic as anomalous if it deviates from the baseline behavior. They can also be more complex and resource-intensive to implement, requiring extensive training data and processing power to build accurate models of network behavior (Alazab et al., 2012).

Overall, AIDS can be an effective complement to SIDS in providing comprehensive intrusion detection and prevention in cybersecurity. By combining both approaches, organizations can achieve a more comprehensive and effective defense against cyber threats.

9.5 ATTACKS ON IOT ECOSYSTEM

IoT technology incorporates a variety of devices, such as sensors and processors, to enable efficient data sharing and network connectivity. However, as the number of devices connected to the network increases, data security becomes a major concern. IoT security aims to safeguard the information transmitted across various IoT devices and networks. Without adequate security measures in place, the loss of data can significantly impact various industries and individuals, leading to catastrophic consequences (Khraisat et al., 2019).

IoT has captured the attention of individuals and organizations across multiple sectors, offering numerous benefits. However, as IoT continues to expand, several security issues have emerged, leading to attacks that have impeded the development of new IoT applications. This report section focuses on IoT security, its challenges, impacts, and various types of IoT attacks. Therefore, IoT security measures must address concerns regarding security, privacy, and confidentiality to safeguard against these attacks. For instance, an attack on traffic lights or autonomous vehicles could result in chaos, increased pollution, and even physical harm, including severe collisions leading to injuries. The internet enables various devices and equipment in homes and offices to be virtually connected, enabling remote monitoring and control of their operations.

Below is a summary of the main reasons why IoT is often targeted by malware:

- All devices in an IoT system need to remain constantly powered on, making it easy for attackers to identify and exploit vulnerable equipment.
- Devices in an IoT system are interconnected, allowing attackers potential access to multiple devices from a single compromised device.
- Defending against attacks in a network of interconnected devices is more challenging than protecting a single computer due to the complexity and diversity of the system.
- Weak encryption and passwords can make interconnected devices vulnerable to malware attacks.
- IoT devices are always connected to the internet, exposing them to potential threats around the clock. This unlimited connectivity makes them vulnerable to incoming traffic signals, thereby increasing the risk of malware attacks.

Furthermore, the characteristics of malware can differ in a single device versus an IoT environment.

Some of the common attacks on the IoT ecosystem include:

Botnets: Botnets are networks of compromised devices that are controlled by a single attacker. These devices can launch DDoS attacks, spam campaigns, and other malicious activities.

Malware: Malware is malicious software that can infect IoT devices and steal sensitive information, disrupt device functionality, or spy on users.

Man-in-the-Middle (MItM) Attacks: MItM attacks intercept communications between IoT devices and their intended destination, allowing attackers to eavesdrop, modify, or manipulate the communication.

Denial-of-Service (DoS) Attacks: DoS attacks aim to overwhelm an IoT device or network with traffic, making it unavailable to users.

Physical Attacks: Physical attacks involve physically accessing and tampering with IoT devices or infrastructure, such as cutting wires or stealing devices.

Password Attacks: Password attacks involve guessing or cracking passwords to gain unauthorized access to IoT devices or networks.

To prevent these attacks, it is essential to secure IoT devices with strong passwords, keep software up to date, use encryption, and restrict access to devices to authorized users only. Regular vulnerability assessments and penetration testing are also important to identify and mitigate potential security risks. IoT systems are typically designed with multiple layers, each serving a specific function in the overall architecture. These layers provide a framework for understanding the different areas of vulnerability in an IoT system. Below are the common layers in an IoT system architecture and the potential attacks that can occur in each layer:

> *Device Layer*: This is the bottom layer of an IoT system and includes the physical devices themselves. Attacks on this layer can include physical attacks, such as tampering with the hardware, or malware attacks that target the firmware or software running on the device.
>
> *Connectivity Layer*: This layer includes the communication protocols and networks that connect IoT devices and the cloud. Security threats directed toward this layer involve various types of attacks, including MItM attacks, where an attacker intercepts and modifies communication between devices, or DoS attacks, where an attacker floods the network with traffic to disrupt communication.
>
> *Cloud Layer*: This layer includes the cloud infrastructure that IoT devices use to store and process data. Attacks on this layer can include data breaches, where an attacker gains unauthorized access to data stored in the cloud, or DoS attacks that overwhelm cloud infrastructure and render it unavailable.
>
> *Application Layer*: This layer includes the software applications that allow users to interact with IoT devices. Attacks on this layer can include phishing attacks, where an attacker sends a fraudulent message to a user to obtain their login credentials or other sensitive information, or software attacks that exploit vulnerabilities in the application code.
>
> *Business Layer*: This layer includes the business logic and processes that govern the operation of the IoT system. Attacks on this layer can include social engineering attacks that manipulate personnel or stakeholders in the IoT system or supply chain attacks that target third-party vendors or suppliers.

To secure an IoT system, it is important to consider each layer of the system architecture and implement appropriate security measures at each level. This includes using strong authentication and access control measures, encrypting data both in transit and at rest, and regularly updating software and firmware to patch known vulnerabilities. Additionally, continuous monitoring and analysis of IoT systems can help detect and respond to potential security threats before they can cause significant harm, as depicted in Figure 9.1.

The three fundamental layers of an IoT system are the perception layer, network layer, and application layer (Figure 9.2).

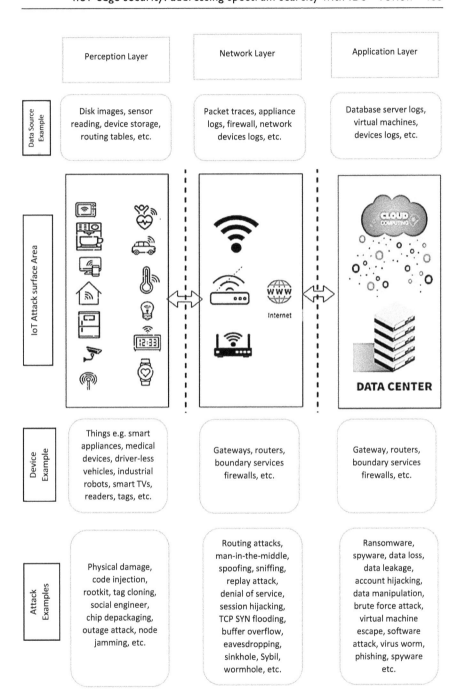

Figure 9.1 IoT system architecture with layers where attacks can occur (Khraisat & Alazab, 2021)

Attribute	Personal Computer	Internet of Things
Execution Platform Heterogeneity	Low	High
Malware family Variety	High	Low
Intrusion Detection Technique	Easy	Difficult
Internal Analysis	Easy	Very Difficult
Sandbox Execution	Easy	Difficult
Removing Malware	Medium	Very Difficult
Sandbox Execution	Easy	Difficult

Figure 9.2 The types of attacks, as well as how these attacks impact the IoT network and their implications, are described (Khraisat & Alazab, 2021).

Attribute	Personal Computer	Internet of Things
Execution platform heterogeneity	Low	High
Malware family variety	High	Low
Intrusion detection technique	Easy	Difficult
Internal analysis	Easy	Very difficult
Sandbox execution	Easy	Difficult
Removing malware	Medium	Very difficult
Sandbox execution	Easy	Difficult

9.5.1 Physical/perception layer

In an IoT system, the physical/perception layer refers to the layer where sensors, actuators, and other physical devices are located. This layer is responsible for collecting and processing data from the physical environment and providing feedback to the system. The physical/perception layer is crucial to the overall functionality of an IoT system. However, it is also a potential area of vulnerability. Attacks on this layer can include physical attacks, such as tampering with sensors or disrupting communication channels between devices (Haque, Ahmad, John, et al., 2023; Haque, Ahmad, Sonal, et al., 2023). For example, an attacker might physically alter a temperature sensor to provide false readings, or they might disrupt communication between sensors and actuators to cause a malfunction in the system.

However, this layer is also vulnerable to several security threats, some of which are:

1. *Tampering*: Attackers may physically tamper with the sensors or the data they generate to manipulate the system's functioning or obtain unauthorized access to the network.
2. *Spoofing*: This refers to the act of creating fake sensor data that can trick the system into taking unwanted actions or making wrong decisions.
3. *Eavesdropping*: Attackers may intercept the communication between the sensors and the network to gain access to sensitive information.
4. *Denial-of-Service (DoS) Attacks*: Attackers may flood the sensors with a large volume of data or cause them to malfunction, thereby disrupting the entire IoT system's functioning.
5. *Physical Destruction*: An attacker may physically damage the sensors, rendering them inoperable and causing significant damage to the system.
6. *Energy Depletion*: Low-power sensors can be vulnerable to attacks that drain their battery power, leading to their malfunctioning.
7. *Environmental Factors*: Physical sensors may be susceptible to various environmental factors such as temperature, humidity, and electromagnetic interference, which may cause them to malfunction or generate inaccurate data.

   To secure the physical/perception layer of an IoT system, it is important to implement physical security measures, such as tamper-resistant hardware and secure communication protocols. It is also important to ensure that the sensors and other physical devices are properly calibrated and regularly maintained to prevent malfunctions. Additionally, privacy and data protection measures, such as encryption and access controls, should be implemented to protect the data collected and processed in this layer.

9.5.2 Software/application layer attacks in IoT

Software/application layer attacks in IoT refer to attacks that target the software and applications running on top of the IoT infrastructure (ÖZALP et al., 2022). These attacks can exploit vulnerabilities in the software and applications, potentially leading to data theft, system malfunctions, or other security breaches. Examples of software/application layer attacks in IoT include:

1. *Malware Attacks*: These attacks involve an attacker infecting a device or network with malware, such as viruses, worms, or Trojan horses. The goal of such an attack is to gain control of the system, steal data, or cause a malfunction.

2. *Denial-of-Service (DoS) Attacks*: These attacks involve an attacker flooding a device or network with traffic, preventing legitimate users from accessing the system. The goal of such an attack is to disrupt the normal functioning of the system or render it inoperable.
3. *Man-in-the-Middle (MitM) Attacks*: The goal of such an attack is to steal sensitive information, such as user credentials, or manipulate the data to cause a malfunction.
4. *SQL Injection Attacks*: These attacks involve an attacker exploiting a vulnerability in a web application to inject malicious SQL code into the application's database. The goal is to steal sensitive data or modify the database to cause a malfunction.
5. *Cross-site Scripting (XSS) Attacks*: These attacks involve an attacker injecting malicious code into a web page viewed by other users. The goal is to steal sensitive data or gain control of the user's device.

To prevent software/application layer attacks in IoT, regular software updates and patching can help prevent vulnerabilities from being exploited. Additionally, user education and awareness can help prevent attacks that rely on social engineering or other forms of manipulation.

9.5.3 Network layer attacks in IoT

Network layer attacks in IoT refer to attacks that target the communication infrastructure used by IoT devices to exchange data (Alimul Haque, Haque, Rahman, Kumar, 2022). These attacks can exploit vulnerabilities in the network protocols and infrastructure, potentially leading to data theft, system malfunctions, or other security breaches. Examples of network layer attacks in IoT include:

1. *Man-in-the-Middle (MitM) Attacks*: The goal of such an attack is to steal sensitive information, such as user credentials, or manipulate the data to cause a malfunction.
2. *Denial-of-Service (DoS) Attacks*: These attacks involve an attacker flooding a network with traffic, preventing legitimate users from accessing the system. The goal of such an attack is to disrupt the normal functioning of the network or render it inoperable.
3. *Spoofing Attacks*: These attacks involve an attacker masquerading as a legitimate device or network resource. The goal of such an attack is to gain unauthorized access to the network or steal sensitive data.
4. *Routing Attacks*: These attacks involve an attacker manipulating the routing tables used by IoT devices to send and receive data. The goal of such an attack is to intercept or modify data transmitted between devices in the network.
5. *Botnet Attacks*: These attacks involve an attacker using a network of compromised devices to launch coordinated attacks on other devices

or networks. The goal of such an attack is to disrupt the normal functioning of the network or steal sensitive data.

6. *Attacks on RPL*: Routing Protocol for Low-Power and Lossy Networks(RPL) is a popular routing protocol used in many IoT networks to facilitate communication between devices. However, like any protocol, RPL is susceptible to attacks that can compromise the security and stability of the network. Here are some common attacks that can be launched against RPL:

   Rank Attacks: In an RPL network, each device is assigned a rank based on its position in the network topology. Attackers can launch rank attacks to manipulate the ranking of devices, allowing them to gain control of the network or redirect traffic to malicious nodes.

   Sybil Attacks: In an RPL network, a Sybil attack can be used to manipulate the routing tables, redirect traffic, or cause nodes to fail.

   Sinkhole Attacks: In a sinkhole attack, an attacker creates a malicious node that advertises a low hop count to attract traffic to it. In an RPL network, a sinkhole attack can be used to redirect traffic to the malicious node, allowing the attacker to intercept and manipulate data.

   Blackhole Attacks: In a blackhole attack, an attacker creates a malicious node that drops all traffic it receives, effectively blocking communication in the network. In an RPL network, a blackhole attack can disrupt communication between devices or prevent specific nodes from communicating with each other.

   Wormhole Attacks: In an RPL network, a wormhole attack can redirect traffic, intercept data, or cause nodes to fail.

To prevent attacks on RPL, it is important to implement strong security measures, such as encryption, authentication, and access control. Regular network monitoring and testing can also help identify potential vulnerabilities and prevent attacks before they can cause significant harm. Additionally, device firmware updates can help patch any known vulnerabilities in the RPL implementation. To evade network layer attacks in IoT, it is important to implement strong network security measures, such as firewalls, IDS, and access controls. Secure communication protocols, such as Transport Layer Security (TLS) or Internet Protocol Security (IPsec), can help prevent data interception and manipulation. Regular network monitoring and testing can also help identify potential vulnerabilities and prevent attacks before they can cause significant harm.

9.6 INTRUSION DETECTION DATASETS

Intrusion detection datasets are essential resources for the development, training, and evaluation of IDS. These datasets contain a collection of network traffic data that includes benign traffic and malicious attacks. IDS developers use these datasets to train their models and test their effectiveness in detecting and responding to various types of attacks.

KDD Cup 99 is a popular intrusion detection dataset involved in several cutting-edge technology projects, including the development of the internet, GPS, and cybersecurity. One of the notable projects undertaken by DARPA is the KDD Cup99 competition, which aimed to improve the detection of network intrusion attacks.

This benchmark dataset simulated network traffic in a military environment, comprising approximately 5 million Transmission Control Protocol/Internet Protocol (TCP/IP) connection records, divided into training and testing sets. The goal of the competition was to develop an efficient and accurate algorithm to classify network connections as either normal or attack (ÖZALP et al., 2022).

9.6.1 CAIDA (center for applied internet data analysis)

CAIDA is a research group that collects and analyzes internet data to improve the security and resilience of the internet. CAIDA has also provided several datasets commonly used for building and comparatively evaluating IDS (Hick et al., 2007). Some of the popular datasets provided by CAIDA are discussed below, along with their features and limitations:

CAIDA Anonymized 2016: This dataset contains anonymized traffic data collected from a regional ISP in North America for one week in 2016. The dataset consists of several types of traffic, including TCP, User Datagram Protocol (UDP), Internet Control Messaging Protocol (ICMP), and Domain Name Server (DNS) traffic. The dataset is useful for studying the characteristics of network traffic and detecting anomalies. However, the dataset is relatively small, and the data are anonymized, which may limit the accuracy of the analysis.

CAIDA-DDoS-2018: This dataset contains traffic data collected during a DDoS attack on a research network in 2018. The dataset includes various types of traffic, including TCP, UDP, and ICMP traffic. The dataset is useful for studying the characteristics of DDoS attacks and developing effective countermeasures. However, the dataset is relatively small and may not reflect the characteristics of large-scale DDoS attacks.

CAIDA 2018 Dataset: This dataset contains traffic data collected from multiple locations around the world during 2018. The dataset includes several types of traffic, including TCP, UDP, ICMP, and DNS traffic. The dataset is useful for studying the characteristics of network traffic and detecting anomalies. However, the dataset is relatively large, which may require significant computational resources for analysis.

CAIDA Intrusion Detection Evaluation Dataset: This dataset contains network traffic data and simulated attacks collected in a controlled environment. The dataset includes various types of attacks, such as buffer overflow, SQL injection, and cross-site scripting attacks. The dataset is useful for evaluating the performance of IDS systems, but the simulated attacks may not reflect real-world attacks accurately.

Overall, the CAIDA datasets are useful resources for studying the characteristics of network traffic and evaluating the performance of IDS systems. However, each dataset has its own features and limitations that researchers and practitioners must consider when selecting a dataset for analysis.

NSL-KDD: NSL-KDD is a modified version of the KDD Cup 99 dataset that addresses some of the limitations of the original dataset. It includes new attack scenarios, additional features, and a reduced number of redundant records. NSL-KDD contains 41 features and four types of attacks, including DoS, Probe, and R2L (Tavallaee et al., 2009).

CICIDS2017: The Canadian Institute for Cybersecurity Intrusion Detection Dataset (CICIDS) 2017 is a recent dataset that includes realistic attack scenarios and modern IoT devices. The dataset includes both network traffic data and system logs, with 15 types of attacks, such as DoS, PortScan, and Botnet.

9.6.2 ADFA-LD and ADFA-WD intrusion detection datasets

ADFA Intrusion Detection Datasets consist of two datasets, Australian Defence Force Academy Linux Database (ADFA-LD) and Australian Defence Force Academy-Windows Dataset (ADFA-WD), created by the Network Research Group at ADFA for developing and evaluating IDS. ADFA-LD (Log-based Dataset) contains network traffic data collected from a simulated network environment and system logs from a Linux-based operating system. The dataset includes benign traffic and four types of attacks: DoS, Probe, R2L, and U2R. It contains approximately 5.5 million network packets and over 300,000 system log entries. ADFA-WD (Windows-based Dataset) contains network traffic data and system logs from a Windows-based operating system. The dataset includes benign traffic and four types of attacks, including DoS, Probe, R2L, and U2R. It contains approximately 80,000 network packets and over 7,000 system log entries.

Both datasets include features such as source IP address, destination IP address, protocol type, service, and duration. The datasets also include metadata such as timestamps, packet size, and other features that are used for intrusion detection. ADFA-LD and ADFA-WD datasets have several advantages, such as being created in a controlled environment, containing real-world scenarios, and including both network traffic data and system logs (Xie et al., 2014). However, the datasets also have some limitations, such as being relatively small compared to other intrusion detection datasets and not containing a wide range of attacks. Overall, ADFA-LD and ADFA-WD are valuable datasets for developing and evaluating IDS, and they have been used in various research studies. However, it is important to consider the limitations of these datasets and use them in conjunction with other datasets to ensure a comprehensive evaluation of IDS (Sharafaldin et al., 2018).

9.6.3 ISCX (information security center of excellence) 2012

The ISCX 2012 dataset has several features and limitations that researchers and practitioners must consider when selecting a dataset for analysis. Some of the notable features and limitations of the dataset are discussed below (Shiravi et al., 2012):

Large and Diverse Dataset: The ISCX 2012 dataset is large and contains a diverse range of network traffic and attacks, making it a useful resource for studying the characteristics of network traffic and developing effective IDS.

Real-world Traffic: The dataset was collected from a real-world university campus network, making it more representative of actual network traffic than simulated datasets.

Pre-processed: The dataset has been pre-processed to remove duplicate packets and perform other data cleaning tasks, which may save time in data preparation.

9.6.3.1 Limitations

Limited Timeframe: The dataset covers only four months of network traffic, which may not be sufficient to capture the full range of network traffic and attacks.

Limited Scope: The dataset was collected from a single university campus network, which may not reflect the characteristics of other networks, such as enterprise or government networks.

Lack of Ground Truth Labels: The dataset does not include ground truth labels for some types of attacks, making it difficult to evaluate the performance of IDS systems accurately.

Overall, the ISCX 2012 dataset is a useful resource for building and evaluating IDS systems, but researchers and practitioners must consider the dataset's features and limitations when selecting a dataset for analysis (Shiravi et al., 2012).

9.6.4 IoT botnet

The Bot-IoT dataset is a publicly available dataset consisting of network traffic data captured from a simulated IoT environment. Created by researchers at the University of New Brunswick, it is intended for developing and evaluating IDS for IoT networks (Haque et al., 2020). The dataset contains network traffic data collected from a simulated IoT network consisting of several IoT devices, including cameras, routers, and smart light bulbs. The data was collected over a period of several weeks and includes both normal network traffic and traffic generated by a number of different types of botnets. The Bot-IoT dataset is one of the few publicly available datasets specifically designed for IoT security research, and it has been used in numerous research studies focused on developing and evaluating IDS for IoT networks. It is available for download from the University of New Brunswick website (Koroniotis et al., 2018).

These datasets play a crucial role in developing and evaluating IDS. However, there are some limitations to these datasets. For example, some datasets may not represent real-world scenarios accurately, and attacks in the datasets may be outdated or not representative of the current threat landscape. It is important to consider these limitations while selecting and using intrusion detection datasets.

9.6.4.1 Comparison of public IDS datasets

Machine learning techniques are utilized in intrusion detection, and the datasets employed for these techniques play a crucial role in their realistic evaluation. The characteristics of the datasets are summarized in Table 10. The KDD'99 dataset, which is widely used for wired network environments, and similar datasets, are not suitable for developing optimized IDS targeting IoT ecosystems.

9.7 CHALLENGES OF IOT IDS

An IDS is a security tool that monitors network traffic for potential security breaches or malicious activity (Whig et al., 2022). In the context of IoT, IDS can detect and respond to attacks targeting IoT devices and the network infrastructure.

However, IoT poses several unique challenges for IDS, including:

Scale: The number of devices in an IoT network can range from a few to millions, and each device generates a significant amount of data. This large volume of data can make it challenging for IDS to process and analyze data in real time.

Heterogeneity: IoT networks can comprise devices from different vendors, with varying hardware, software, and communication protocols. This heterogeneity makes it challenging for IDS to identify and classify devices and their associated traffic.

Dynamic Topology: IoT networks can be highly dynamic, with devices joining and leaving the network frequently. This dynamic topology can make it challenging for IDS to maintain an accurate map of the network and detect potential attacks in real time.

Limited Resources: Many IoT devices have limited processing power, memory, and battery life. IDS that require significant computational resources may not be suitable for such devices.

Encrypted Traffic: IoT devices often use encrypted communication protocols, which pose a challenge for IDS to analyze the content of the traffic and detect potential threats.

To overcome these challenges, researchers are developing new techniques and tools specifically designed for IoT IDS. For example, lightweight IDS that use machine learning and statistical techniques to detect anomalies in network traffic can help overcome the resource limitations of IoT devices. Hybrid IDS, which combine both signature-based and anomaly-based detection methods, can improve detection accuracy and reduce false positives.

9.7.1 Challenge of IoT IDS on intrusion evasion detection

IDS are essential tools in cybersecurity for detecting and responding to cyber threats. With the rise of IoT, an increasing need for IDS that can handle the unique challenges posed by IoT devices is felt. One of the biggest challenges of IoT IDS is intrusion evasion detection (Jha et al., 2022).

Intrusion evasion is a technique attackers employ to bypass IDS and evade detection. This technique involves modifying the behavior of attacks to avoid triggering IDS rules and signatures. Intrusion evasion can be particularly effective in IoT, where devices may have limited processing power and memory, making it difficult to run complex IDS algorithms. To address the challenge of intrusion evasion in IoT IDS, researchers are exploring various techniques, such as:

Anomaly Detection: Anomaly detection is a machine learning technique that can detect deviations from normal behavior. By training anomaly detection algorithms on normal device behavior, IoT IDS can detect and respond to abnormal behavior that may indicate an attack.

Traffic Analysis: Traffic analysis involves monitoring network traffic to identify patterns and anomalies. By analyzing network traffic in real time, IoT IDS can detect and respond to suspicious behavior that may indicate an attack.

Dynamic Rule Generation: Dynamic rule generation involves creating IDS rules on the fly based on the behavior of IoT devices. By adapting IDS rules to the specific behavior of each device, IoT IDS can detect and respond to attacks that traditional IDS rules may miss.

Data Fusion: Data fusion involves combining data from multiple sources, such as sensors and network traffic, to detect and respond to attacks.

By using data from multiple sources, IoT IDS can improve the accuracy of intrusion detection and reduce the risk of false positives and false negatives.

Overall, addressing the challenge of intrusion evasion detection in IoT IDS requires innovative solutions that can handle the unique challenges posed by IoT devices. By leveraging machine learning techniques and data fusion, IoT IDS can improve their ability to detect and respond to cyber threats in real time, thereby reducing the risk of data breaches and other security incidents (Ahmad & Afzal, n.d.).

9.8 DISCUSSION AND CONCLUSION

This chapter presents a comprehensive analysis of intrusion detection system methodologies, deployment and validation strategies, datasets, and technologies concerning IoT. The advantages and limitations of these technologies are discussed in detail. Various IDS for identifying IoT attacks are reviewed; however, due to the intricate IoT architecture, these approaches may not detect all attacks. Recent research findings are summarized, and contemporary models for improving the performance of IoT IDS to tackle security concerns are explored. Furthermore, attention is drawn to the inadequacies of conventional IoT IDS, along with examining existing IDS challenges and potential future research directions. To establish a trustworthy IoT IDS that caters to various device categories, designing a new IDS is crucial. Four crucial constituents are pinpointed as indispensable for constructing a dependable IoT IDS. Firstly, the IDS must be highly adaptable to unconventional IoT communication systems, given that IoT sensors may display normal behavior that is, in fact, malicious. Secondly, it should be equipped to detect zero-day attacks as new vulnerabilities are identified. Thirdly, it should be an autonomous IDS that leverages modern machine learning and deep learning techniques to learn from extensive IoT data. This review aims to assist researchers interested in developing new IDS to address IoT security concerns within the context of IoT communication.

REFERENCES

Abbasi, A., Wetzels, J., Bokslag, W., Zambon, E., & Etalle, S. (2014). On emulation-based network intrusion detection systems. *Research in Attacks, Intrusions and Defenses: 17th International Symposium, RAID 2014, Gothenburg, Sweden, September 17–19, 2014. Proceedings 17*, 384–404.

Aburomman, A. A., & Reaz, M. B. I. (2017). A survey of intrusion detection systems based on ensemble and hybrid classifiers. *Computers & Security*, *65*, 135–152.

Agrawal, S., & Agrawal, J. (2015). Survey on anomaly detection using data mining techniques. *Procedia Computer Science*, *60*, 708–713.

Ahmad, S., & Afzal, M. M. (n.d.). A study and survey of security and privacy issues in cloud computing. *International Journal of Engineering Research & Technology (IJERT), 50,* 1009–1014.

Ahmad, S., Jha, S., Abdeljaber, H. A. M., Imam Rahmani, M. K., Waris, M. M., Singh, A., & Yaseen, M. (2022). An integration of IoT, IoC, and IoE towards building a green society. *Scientific Programming, 2022.* Wiley, Hoboken, NJ.

Ahmad, S., Jha, S., Alam, A., Yaseen, M., & Abdeljaber, H. A. M. (2022). A novel AI-based stock market prediction using machine learning algorithm. *Scientific Programming, 2022.* Wiley, Hoboken, NJ.

Alimul Haque, M., Haque, S., Rahman, M., Kumar K., & Zeba, S. (2022). Potential applications of the Internet of Things in sustainable rural development in India. *Proceedings of Third International Conference on Sustainable Computing, 1404,* 455–467. https://doi.org/10.1007/978-981-16-4538-9_45

Annachhatre, C., Austin, T. H., & Stamp, M. (2015). Hidden Markov models for malware classification. *Journal of Computer Virology and Hacking Techniques, 11,* 59–73.

Benkhelifa, E., Welsh, T., & Hamouda, W. (2018). A critical review of practices and challenges in intrusion detection systems for IoT: Toward universal and resilient systems. *IEEE Communications Surveys & Tutorials, 20*(4), 3496–3509.

Bzai, J., Alam, F., Dhafer, A., Bojovi, M., Altowaijri, S. M., Khan Niazi, I., & Mehmood, R. (2022). Machine learning-enabled Internet of Things (IoT): Data, applications, and industry perspective. *Electronics, 11*(17), 2676. https://doi.org/10.3390/electronics11172676

Chaabouni, N., Mosbah, M., Zemmari, A., Sauvignac, C., & Faruki, P. (2019). Network intrusion detection for IoT security based on learning techniques. *IEEE Communications Surveys & Tutorials, 21*(3), 2671–2701.

Granjal, J., Monteiro, E., & Silva, J. S. (2015). Security in the integration of low-power wireless sensor networks with the internet: A survey. *Ad Hoc Networks, 24,* 264–287.

Haque, M. A., Sonal, D., Haque, S., Nezami, M. M., & Kumar, K. (2020). An IoT-based model for defending against the novel coronavirus (COVID-19) outbreak. *Solid State Technology, 29,* 592–600.

Haque, M. A., Haque, S., Kumar, K., & Singh, N. K. (2021). A comprehensive study of cyber security attacks, classification, and countermeasures in the Internet of Things. In *Digital Transformation and Challenges to Data Security and Privacy* (pp. 63–90). IGI Global, Hershey, PA.

Haque, S. (2021). Blockchain technology for IoT security. *Turkish Journal of Computer and Mathematics Education (TURCOMAT), 12*(7), 549–554.

Haque, S., Zeba, S., Alimul Haque, M., Kumar, K., & Ali Basha, M. P. (2021). An IoT model for securing examinations from malpractices. *Materials Today: Proceedings, 81*(2), 371–376. https://doi.org/10.1016/j.matpr.2021.03.413

Haque, M. A., Sonal, D., Haque, S., Rahman, M., & Kumar, K. (2022). Learning management system empowered by machine learning. *AIP Conference Proceedings, 2393,* 1.

Haque, M. A., Almrezeq, N., Haque, S., & Abd El-Aziz, A. A. (2022). Device access control and key exchange (DACK) protocol for Internet of Things. *International Journal of Cloud Applications and Computing (IJCAC), 12*(1), 1–14.

Haque, M. A., Ahmad, S., John, A., Mishra, K., Mishra, B. K., Kumar, K., & Nazeer, J. (2023). Cybersecurity in universities: An evaluation model. *SN Computer Science, 4*(5), 569. https://doi.org/10.1007/s42979-023-01984-x

Haque, M. A., Ahmad, S., Sonal, D., Abdeljaber, H. A. M., Mishra, B. K., Eljialy, A. E. M., Alanazi, S., & Nazeer, J. (2023). Achieving organizational effectiveness through machine learning based approaches for malware analysis and detection. *Data and Metadata, 2*, 139.

Haque, M. A., Haque, S., Zeba, S., Kumar, K., Ahmad, S., Rahman, M., Marisennayya, S., & Ahmed, L. (2023). Sustainable and efficient E-learning Internet of Things system through blockchain technology. *E-Learning and Digital Media, 2023*, 1–20. https://doi.org/10.1177/20427530231156711

Haque, S., Haque, M. A., Kumar, D., Mishra, K., Islam, F., Ahmad, S., Kumar, K., & Mishra, B. K. (2023). Assessing the impact of IoT enabled E-learning system for higher education. *SN Computer Science, 4*(5), 459.

Hindy, H., Brosset, D., Bayne, E., Seeam, A., Tachtatzis, C., Atkinson, R., & Bellekens, X. (2018). *A taxonomy and survey of intrusion detection system design techniques, network threats and datasets.* Preprint/Working Paper. arXiv.org, Ithaca, NY.

Jha, S., Routray, S., & Ahmad, S. (2022). An expert system-based IoT system for minimisation of air pollution in developing countries. *International Journal of Computer Applications in Technology, 68*(3), 277–285.

Khraisat, A., Gondal, I., Vamplew, P., & Kamruzzaman, J. (2019). Survey of intrusion detection systems: Techniques, datasets and challenges. *Cybersecurity, 2*(1), 1–22.

Khraisat, A., & Alazab, A. (2021). A critical review of intrusion detection systems in the internet of things: Techniques, deployment strategy, validation strategy, attacks, public datasets and challenges. *Cybersecurity, 4*, 1–27.

Koroniotis, N., Moustafa, N., Sitnikova, E., & Turnbull, B. (2018). The Bot-IoT Dataset. *UNSW Canberra at ADFA.* https://Research.Unsw.Edu.Au/Projects/Bot-Iot-Dataset (Accessed Jun. 02, 2021).

Koroniotis, N., Moustafa, N., Sitnikova, E., & Turnbull, B. (2019). Towards the development of realistic botnet dataset in the Internet of Things for network forensic analytics: Bot-iot dataset. *Future Generation Computer Systems, 100*, 779–796.

Kreibich, C., & Crowcroft, J. (2004). Honeycomb: Creating intrusion detection signatures using honeypots. *ACM SIGCOMM Computer Communication Review, 34*(1), 51–56.

Masud, M., Muhammad, G., Alhumyani, H., Alshamrani, S. S., Cheikhrouhou, O., Ibrahim, S., & Hossain, M. S. (2020). Deep learning-based intelligent face recognition in IoT-cloud environment. *Computer Communications, 152*, 215–222. https://doi.org/10.1016/j.comcom.2020.01.050

Özalp, A. N., Albayrak, Z., Çakmak, M., & Özdoğan, E. (2022). Layer-based examination of cyber-attacks in IoT. In *2022 International Congress on Human-Computer Interaction, Optimization and Robotic Applications (HORA)*, Ankara, Turkey (pp. 1–10).

Qinxia, H., Nazir, S., Li, M., Ullah Khan, H., Lianlian, W., & Ahmad, S. (2021). AI-enabled sensing and decision-making for IoT systems. *Complexity, 2021*.

Santos, L., Rabadao, C., & Gonçalves, R. (2018). Intrusion detection systems in Internet of Things: A literature review. In *2018 13th Iberian Conference on Information Systems and Technologies (CISTI)*, Caceres, Spain (pp. 1–7).

Sharafaldin, I., Lashkari, A. H., & Ghorbani, A. A. (2018). Toward generating a new intrusion detection dataset and intrusion traffic characterization. *ICISSP, 1*, 108–116.

Shiravi, A., Shiravi, H., Tavallaee, M., & Ghorbani, A. A. (2012). Toward developing a systematic approach to generate benchmark datasets for intrusion detection. *Computers & Security, 31*(3), 357–374.

Srivastava, J., Routray, S., Ahmad, S., & Waris, M. M. (2022). Internet of Medical Things (IoMT)-based smart healthcare system: Trends and progress. *Computational Intelligence and Neuroscience, 2022,* 9768292.

Symantec, O. (2017). "Dragonfly: Western energy sector targeted by sophisticated attack group."

Tavallaee, M., Bagheri, E., Lu, W., & Ghorbani, A. A. (2009). A detailed analysis of the KDD CUP 99 data set. In *2009 IEEE Symposium on Computational Intelligence for Security and Defense Applications,* Ottawa, ON, Canada (pp. 1–6).

Whig, V., Othman, B., Haque, M. A., Gehlot, A., Qamar, S., & Singh, J. (2022). An empirical analysis of artificial intelligence (AI) as a growth engine for the healthcare sector. In *2022 2nd International Conference on Advance Computing and Innovative Technologies in Engineering (ICACITE),* Greater Noida, India (pp. 2454–2457).

Xie, M., Hu, J., Yu, X., & Chang, E. (2014). Evaluating host-based anomaly detection systems: Application of the frequency-based algorithms to ADFA-LD. *Network and System Security: 8th International Conference, NSS 2014,* Xi'an, China, October 15–17, 2014, *Proceedings 8,* 542–549.

Zarpelão, B. B., Miani, R. S., Kawakani, C. T., & de Alvarenga, S. C. (2017). A survey of intrusion detection in Internet of Things. *Journal of Network and Computer Applications, 84,* 25–37.

Chapter 10

AI systems' security issues
Case study on data breaches

Nthatisi Magaret Hlapisi, Nidhi Sagarwal,
Rachit Garg, and Sudan Jha

10.1 INTRODUCTION

Industrial IoT, or Industrial Internet of Things (IIoT), refers to the integration of internet-connected devices and sensors with industrial systems and processes. It involves the use of networked devices, sensors, and software to collect and exchange data to enhance operational efficiency, improve decision-making, and optimize industrial processes [1,2].

IIoT is applied across various sectors such as manufacturing, energy, transportation, agriculture, and healthcare. IIoT devices are typically embedded with sensors and actuators to monitor and control physical processes or assets. These devices collect data on factors like temperature, pressure, humidity, machine performance, and energy consumption, which is then transmitted to a central system or cloud-based platform for storage, analysis, and visualization. By using advanced analytics and machine learning algorithms, valuable information and knowledge can be extracted from the data collected by IIoT devices. These insights can achieve various objectives, such as predictive maintenance (to predict when industrial equipment or machinery is likely to fail or require maintenance and help prevent unexpected breakdowns and machine downtime), process optimization (improving their productivity and optimizing operations through data analysis), quality control (detecting patterns indicating potential quality issues in production processes), and resource management (optimizing resources by analyzing data related to energy consumption, material usage/wastage).[1–3].

While Industrial IoT provides numerous benefits to industries, such as increased automation, improved efficiency, reduced downtime, enhanced safety, and cost savings, it also presents new challenges, including data security, interoperability, scalability, and privacy concerns.

IIoT is a revolutionary technology reshaping our interaction with the physical world. IIoT devices are becoming more and more prevalent in industrial environments. These devices are intricate systems comprising various elements such as sensors, actuators, controllers, and communication networks. They are frequently deployed in challenging and distant locations, exposing them to physical risks, environmental dangers, and cyber threats. IIoT devices encounter several significant security challenges, which include a lack of standardization, limited resources, a lack of security by design, the complexity of systems, connectivity, and a lack of firmware updates. These have been discussed in detail in the *next section*.

The threat landscape for IIoT devices is constantly evolving, with new threats emerging constantly. Some of the most common threats facing IIoT devices include:

1. *Malware Attacks*: Malware can infect IIoT devices and compromise their functionality, steal sensitive data, or use them as part of a larger botnet.
2. *Physical Attacks*: IIoT devices located in public or unsecured areas can be physically accessed and tampered with, leading to unauthorized access or damage to the device.
3. *Network Attacks*: IIoT devices connected to the internet can be targeted by hackers using various techniques to exploit vulnerabilities in the device's network stack or communication protocols.
4. *Data Breaches*: IIoT devices that collect and transmit sensitive data can be targeted by hackers who attempt to steal this data for financial gain or to gain a competitive advantage.
5. *Supply Chain Attacks*: IIoT devices that untrusted vendors manufacture or distribute can be compromised at the source, leading to security vulnerabilities or backdoors that are difficult to detect.

10.2 OVERVIEW OF SECURITY CHALLENGES IN IIOT DEVICES

IIoT is transforming the way we interact with the physical world, enabling real-time monitoring and control of complex industrial processes. IIoT devices are becoming increasingly prevalent in industrial settings, such as manufacturing, energy, transportation, and agriculture, where they are used to automate processes, monitor performance, and optimize efficiency. However, with the widespread adoption of IIoT devices, new security emerging security challenges must be addressed to ensure the safety and reliability of critical infrastructure (Figure 10.1).

IIoT devices are complex systems consisting of multiple components, including sensors, actuators, controllers, and communication networks. These devices are often deployed in harsh and remote environments, where they are exposed to physical and environmental hazards, as well as cyber threats. Some of the key security challenges facing IIoT devices include:

1. Lack of standardization [2–4].
   IIoT devices are highly diverse and come from many different vendors, making it difficult to establish common security standards and

Figure 10.1 Security challenges in Industrial IoT devices.

best practices. Unlike traditional IT systems, based on well-established standards and protocols, IIoT devices are often proprietary and lack interoperability. This can make it challenging to implement security measures across multiple IIoT devices and ensure their compatibility with existing security infrastructure.

The lack of standardization in IIoT devices can also make it difficult to manage security risks across the entire supply chain. Many IIoT devices are manufactured by third-party vendors, which can introduce security risks if these vendors do not adhere to best practices for security and quality assurance. Furthermore, IIoT devices may be used in conjunction with other devices and systems, which can further complicate security management.

2. Limited resources [5,6]

IIoT devices are often designed with limited resources, such as memory, processing power, and battery life, which makes it challenging to implement advanced security features. Unlike traditional IT systems, which are typically based on powerful servers and desktops, IIoT devices are designed to be small, low-power, and efficient. This can make it difficult to implement security measures such as encryption, authentication, and access control.

The limited resources of IIoT devices can also make it challenging to update their firmware and software to address security vulnerabilities. Firmware updates can require significant processing power and memory, which may not be available on some IIoT devices. In addition, many IIoT devices are deployed in remote or hard-to-reach locations, which can make it difficult to physically update their firmware.

3. Lack of security by design [6–14]

Many IIoT devices are designed with functionality as the primary focus, with security being an afterthought. This can lead to vulnerabilities in the device's architecture and software. Unlike traditional IT systems, which are typically designed with security in mind from the outset, IIoT devices are often designed to meet specific functional requirements, with security being a secondary concern.

The lack of security by design can make it challenging to identify and mitigate security risks in IIoT devices. Security vulnerabilities may be introduced at various stages of the device's development, such as during the design, manufacturing, or deployment phases. Furthermore, the lack of security by design can make it challenging to integrate IIoT devices with existing security infrastructure.

4. Complexity of systems [5–7]

IIoT devices are complex systems that involve multiple layers of hardware, software, and communication protocols. This complexity can make it challenging to identify and mitigate security risks. Unlike traditional IT systems, which are typically based on a small number of hardware and software components, IIoT devices may include dozens or even hundreds of components [15–20].

The complexity of IIoT systems can also make it challenging to manage security risks across the entire system.

10.2.1 Attack vectors for IIoT devices

Attack vectors for IIoT devices refer to the different methods or pathways through which malicious actors can target and exploit vulnerabilities in these devices. These attack vectors can be seen as potential entry points for hackers to gain unauthorized access, manipulate data, disrupt operations, or compromise the security of industrial systems.

Some common attack vectors [3,21–26] for IIoT devices include:

1. *Network-Based Attacks*: Hackers can exploit vulnerabilities in network protocols or devices to gain unauthorized access to IIoT devices or intercept and manipulate data transmissions.
2. *Physical Attacks*: Attackers can physically tamper with IIoT devices, either by stealing them, altering their components, or compromising their physical connections. This can lead to unauthorized access, data theft, or disruption of industrial operations.
3. *Firmware and Software Attacks*: Attackers may target the firmware or software running on IIoT devices by exploiting vulnerabilities or injecting malicious code. This can allow them to gain control over the device, manipulate its behavior, or extract sensitive information.
4. *Social Engineering*: Attackers can employ various social engineering techniques, such as phishing or impersonation, to deceive users or administrators into revealing sensitive information, granting unauthorized access, or downloading malicious software.
5. *Supply Chain Attacks*: Malicious actors may target the supply chain of IIoT devices, compromising them during manufacturing, distribution,

or installation. This can involve introducing backdoors, malware, or counterfeit components into the devices, leading to security breaches.

6. *Insider Threats*: Employees or individuals with authorized access to IIoT devices can intentionally or unintentionally misuse their privileges, leading to unauthorized access, data breaches, or sabotage.

Understanding these attack vectors is crucial for implementing effective security measures and mitigating risks associated with IIoT devices.

Common vulnerabilities in IIoT devices can expose industrial systems to various security risks and compromise the integrity, availability, and confidentiality of data and operations. Some of the most prevalent vulnerabilities include (Figure 10.2):

i. *Weak Authentication and Authorization*: IIoT devices often suffer from weak or default credentials, allowing attackers to easily gain unauthorized access. Inadequate authentication mechanisms make it easier for malicious actors to exploit vulnerabilities and control the devices or access sensitive information. Moreover, insufficient or absent encryption in IIoT devices can result in unauthorized interception and manipulation of data during transmission, making it easier for attackers to compromise the integrity and confidentiality of the information.

ii. *Insecure Communication Protocols*: Many IIoT devices use outdated or insecure communication protocols, leaving them vulnerable to eavesdropping, tampering, and man-in-the-middle attacks. Attackers can intercept or manipulate the data exchanged between devices or between devices and backend systems.

Figure 10.2 Common vulnerabilities in IIoT devices [38].

iii. *Firmware and Software Vulnerabilities*: IIoT devices may contain unpatched software or firmware, making them susceptible to known exploits. Inadequate software development practices and a lack of regular security updates increase the risk of successful attacks.

iv. *Lack of Physical Security*: IIoT devices deployed in industrial settings may not have adequate physical protection. This can result in unauthorized access, tampering, or theft of the devices, allowing attackers to compromise their functionality or extract sensitive information.

v. *Insufficient Monitoring and Logging*: Lack of robust monitoring and logging capabilities in IIoT devices makes it difficult to detect and respond to security incidents effectively. This can delay or hinder the identification of unauthorized activities or anomalies in the system.

vi. *Integration with Legacy Systems*: IIoT devices are often integrated with legacy industrial control systems that may have outdated security measures. Compatibility issues and vulnerabilities in these legacy systems can provide entry points for attackers to exploit and compromise the entire network.

To address these vulnerabilities, organizations should prioritize security by implementing strong authentication mechanisms, encryption, regular firmware updates, secure communication protocols, physical security measures, comprehensive monitoring and logging systems, and effective vulnerability management practices [8,927,28]. Regular security assessments and penetration testing can also help identify and remediate vulnerabilities in IIoT devices before they are exploited by malicious actors.

10.3 THREAT LANDSCAPE FOR IIOT DEVICES

The security challenges facing IIoT devices are diverse and complex, and as they become more interconnected in critical infrastructure, they are attractive targets for malicious actors. To mitigate these risks, organizations must prioritize security and adopt a proactive and holistic approach to IIoT security. By doing so, they can fully realize the benefits of IIoT devices while minimizing the risks of security breaches and disruptions (Figure 10.3).

IIoT devices are vulnerable to unauthorized access, which can be achieved by exploiting device firmware weaknesses or weak authentication mechanisms. Once infiltrated, these devices can launch attacks, tamper with data, or disrupt operations. Robust access control mechanisms and authentication protocols are necessary to counter these threats. Data breaches are also a significant concern, as IIoT devices produce vast amounts of sensitive data that must be protected during transmission and storage. Encryption techniques can mitigate the risk of data breaches. Firmware updates and patch management are equally critical for IIoT device security. Devices running

Risk assessment

Access control

Encryption

Patch management

Physical security

Network security

User training

Figure 10.3 Measures to mitigate security threats in IIoT devices.

on outdated firmware versions are vulnerable to known exploits, so regular updates are necessary to address vulnerabilities and enhance device resilience. However, updating firmware in industrial environments can be challenging as it may interfere with operations, thus requiring careful planning and implementation of update procedures [1–3,21].

Physical security is a crucial aspect of the IIoT threat landscape, as devices deployed in critical infrastructure are vulnerable to tampering, theft, or destruction. Adequate measures like access controls, surveillance systems, and tamper-resistant enclosures are necessary to prevent unauthorized physical access. Network security also plays a vital role in protecting devices from network-based attacks through secure architectures, segmentation, firewalls, and intrusion detection systems. User education and training are essential to address the human element, ensuring users follow best security practices and report suspicious activities. Mitigating IIoT security challenges requires a proactive approach encompassing risk assessment, authentication, encryption, firmware updates, physical security, network security, and user education. Organizations must prioritize security, invest in resilient IIoT systems, and strike a balance between reaping the benefits of IIoT devices and minimizing security risks. By doing so, they can enhance efficiency, productivity, and innovation while avoiding disruptions and breaches [4,21,22].

10.4 SECURITY MEASURES FOR IIOT DEVICES

Securing IIoT devices presents several unique challenges due to the diverse and dynamic nature of these devices, as well as the complex and interconnected systems that they operate within. In this section, we will explore some of the key challenges associated with securing IIoT devices and discuss potential solutions [1–5,21,22].

1. *Device Heterogeneity*:

    One of the biggest challenges in securing IIoT devices is the heterogeneity of these devices. IIoT devices come in a wide range of form factors, operating systems, communication protocols, and other specifications, which can make it difficult to develop a one-size-fits-all security solution. This heterogeneity can also make it difficult to ensure that all IIoT devices are running the latest security patches and firmware updates, which can leave these devices vulnerable to attack.

    To address this challenge, organizations must take a risk-based approach to IIoT security that takes into account the specific characteristics of each device and the risks associated with its use. This may involve implementing different security measures for different types of devices, depending on their level of risk and the resources available for security management.

2. *Resource-constrained Devices*:

    Another challenge in securing IIoT devices is that many of these devices are resource-constrained, meaning that they have limited processing power, memory, and battery life. This can make it difficult to implement complex security measures on these devices, as these measures may consume too many resources and impact device performance.

    To address this challenge, security solutions for IIoT devices must be designed with resource constraints in mind. This may involve implementing lightweight security protocols and encryption algorithms that are optimized for resource-constrained devices. It may also involve using edge computing and other techniques to offload security processing from the device itself and onto more powerful servers or gateways.

3. *Lack of Standardization*:

    The lack of standardization in IIoT devices is another significant challenge in securing these devices. Without common standards for communication protocols, security protocols, and other specifications, it can be difficult to develop interoperable and scalable security solutions for IIoT devices. This lack of standardization can also create a barrier to the adoption of IIoT devices, as organizations may be hesitant to invest in devices that are not interoperable with their existing systems.

    To address this challenge, industry stakeholders must work together to develop common standards for IIoT devices. This may involve developing industry-wide standards for communication protocols, security protocols, and other specifications. It may also involve working with regulatory bodies and other organizations to ensure that these standards are adopted and enforced.

4. *Legacy Systems*:

Many organizations that adopt IIoT devices have existing legacy systems that were not designed with security in mind. These legacy systems may be vulnerable to attack and may also create vulnerabilities in the IIoT devices that are connected to them. For example, a legacy system that is not properly secured may provide an entry point for attackers to gain access to IIoT devices that are connected to it.

To address this challenge, organizations must take a holistic approach to IIoT security that includes securing legacy systems as well as IIoT devices. This may involve implementing security measures such as firewalls and intrusion detection systems to protect legacy systems from attack. It may also involve upgrading legacy systems to more secure platforms that are compatible with IIoT devices.

5. *Lack of Visibility*:

Another challenge in securing IIoT devices is the lack of visibility into the devices and the systems that they are connected to. Many IIoT devices are located in remote or inaccessible locations, which can make it difficult to monitor and manage these devices. In addition, many IIoT devices are connected to proprietary or closed systems, which can make it difficult to gain visibility into the data and traffic that is flowing through these systems.

To address this challenge, organizations must implement solutions that provide visibility into IIoT devices and the systems that they are connected to. This may involve implementing network monitoring and management tools that provide real-time visibility into the traffic flowing through IIoT networks. It may also involve implementing security solutions such as intrusion detection and prevention systems that can help identify and mitigate threats to IIoT devices.

6. *Human Error*:

Finally, one of the biggest challenges in securing IIoT devices is human error. Many security breaches are caused by human error, such as employees clicking on phishing links or failing to follow security protocols. In the context of IIoT devices, human error can lead to a wide range of security vulnerabilities, from misconfigured devices to unsecured network connections.

To address this challenge, organizations must invest in security awareness training for all employees who interact with IIoT devices. This training should cover basic security principles such as password management, phishing awareness, and safe browsing practices. It should also cover specific security protocols and procedures that are relevant to the organization's IIoT devices and systems.

Securing IIoT devices presents several unique challenges, from device heterogeneity to human error. To address these challenges, organizations must take a risk-based approach to IIoT security taking into account the specific characteristics of each device and the risks associated with its use. This may involve implementing different security measures for various types of devices, depending on their level of risk and the resources available for security management.

In addition, organizations must invest in solutions designed with the unique challenges of IIoT devices in mind, such as resource constraints and lack of standardization. This may involve developing lightweight security protocols and encryption algorithms optimized for resource-constrained devices, as well as working with industry stakeholders to develop common standards for IIoT devices.

Ultimately, securing IIoT devices requires a holistic approach that includes not only securing the IIoT devices themselves but also addressing the challenges such as securing legacy systems and providing visibility into the devices and systems they are connected to. By taking a comprehensive approach to IIoT security, organizations can minimize associated risks and ensure they can fully benefit from IIoT while maintaining high levels of security.

10.5 CASE STUDIES

To illustrate the security challenges associated with IIoT devices, this section will discuss several case studies, highlighting the associated risks and threats.

1. *Stuxnet Worm*:

   The Stuxnet worm is a well-known example of a cyberattack targeting Industrial IoT devices. Discovered in 2010, it was specifically designed to target industrial control systems, such as those used in nuclear power plants. The worm spread via infected USB drives and exploited a vulnerability in the Windows operating system to gain access to industrial control systems. Once inside the system, the worm could modify the code of programmable logic controllers (PLCs) to alter the behaviors of industrial processes [29,30].

   The Stuxnet worm was highly sophisticated and demonstrated the potential for cyberattacks to cause physical damage to industrial systems. It also highlighted the importance of strong access controls and authentication mechanisms to prevent unauthorized access to industrial control systems.

2. *Target Data Breach*:

   The Target data breach is another example of a security incident that had significant consequences for industrial organizations. The breach occurred in 2013 and resulted in the theft of 40 million credit card numbers and 70 million customer records [32,33].

The breach was caused by a vulnerability in Target's HVAC system, which was connected to the company's network. Attackers were able to exploit this vulnerability to gain access to Target's network and steal customer data.

The Target breach demonstrated the need for strong access controls and segmentation to prevent unauthorized access to industrial networks and systems. It also highlighted the importance of regular security updates and patches to address known vulnerabilities [31–33].

3. *Mirai Botnet*:

The Mirai botnet is an example of a cyberattack that targeted Internet of Things (IoT) devices, including industrial devices. Discovered in 2016, the botnet compromised a large number of IoT devices, including cameras, routers, and other devices [34,35].

The botnet carried out distributed denial-of-service (DDoS) attacks on a scale never before seen, with some attacks reaching up to 1 terabit per second. These attacks had significant consequences for organizations that relied on IoT devices, including industrial organizations. The Mirai botnet demonstrated the potential for IoT devices to be used as weapons in cyberattacks and highlighted the importance of device hardening and security updates to prevent unauthorized access to IoT devices.

4. *Triton Malware*:

The Triton malware is another example of a cyberattack targeting industrial control systems. Discovered in 2017, the malware was specifically designed to target safety instrumented systems (SIS) used in industrial environments. The malware was able to compromise SIS devices and disable safety systems, which could have had significant consequences for industrial operations. The attack was highly sophisticated and demonstrated the potential for cyberattacks to cause physical harm to industrial workers. [36,37]

The Triton malware highlighted the importance of strong access controls and authentication mechanisms to prevent unauthorized access to industrial control systems. It also demonstrated the need for security monitoring and incident response capabilities to quickly detect and respond to security incidents.

5. *BlackEnergy*:

BlackEnergy is a sophisticated malware that emerged in 2007 and gained attention in 2009. It targets industrial control systems and critical infrastructure and has been used in high-profile cyberattacks, including the 2008 attack on the Ukrainian power grid. It spreads through phishing emails and software vulnerabilities and has capabilities such as remote control and data destruction [23,24].

6. *NotPetya Malware*:

The NotPetya malware is another example of a cyberattack that had significant consequences for industrial organizations. The malware was discovered in 2017 and was able to spread rapidly through industrial networks, causing significant disruption to operations. It exploited a vulnerability in widely used accounting software to gain access to industrial networks. Once inside the network, the malware spread quickly and caused widespread damage. [25]

The NotPetya malware highlighted the importance of regular security updates and patches to address known vulnerabilities. It also demonstrated the need for strong access controls and segmentation to prevent unauthorized access to industrial networks [25,26].

Overall, these case studies demonstrate the significant risks and threats associated with Industrial IoT devices. They highlight the need for strong security measures, including access controls, authentication mechanisms, security monitoring, and incident response capabilities, to prevent unauthorized access and quickly detect and respond to security incidents.

In addition to these measures, it is also important for industrial organizations to stay up-to-date on the latest security threats and vulnerabilities and to implement a proactive approach to security. This may include regular security audits and assessments, penetration testing, and employee training on security best practices.

Industrial organizations should also work closely with their vendors and suppliers to ensure that all devices and software used in their operations are secure and up-to-date. This may include requiring vendors to adhere to strict security standards and conducting regular security assessments of their products.

Finally, it is important for industrial organizations to have a comprehensive disaster recovery and business continuity plan in place to minimize the impact of security incidents on their operations. This plan should include regular backups of critical data and systems, as well as clear procedures for responding to security incidents and restoring operations as quickly as possible.

Securing Industrial IoT devices can be a daunting task due to various challenges. However, there are several best practices that organizations can follow to improve the security posture of their IIoT devices. We will discuss some of these best practices in detail.

1. Implement a risk-based approach [6,7,27]

   As discussed earlier, implementing a risk-based approach is crucial when securing IIoT devices. This approach involves identifying the risks associated with each device and prioritizing security measures accordingly. For example, a device that controls critical infrastructure should be secured with higher priority than one that monitors non-critical equipment.

To implement a risk-based approach, organizations should conduct a thorough risk assessment that includes identifying assets, threats, vulnerabilities, and potential impacts. Based on the results of the risk assessment, organizations can prioritize security measures and allocate resources accordingly.

2. Implement a defense-in-depth strategy [3,21,22,28]

A defense-in-depth strategy involves implementing multiple layers of security controls to protect IIoT devices. This approach is based on the principle that no single security control can provide complete protection against all threats.

To implement a defense-in-depth strategy, organizations should implement a combination of technical and administrative controls. Technical controls may include firewalls, intrusion detection and prevention systems, encryption, and access controls. Administrative controls may include security policies, procedures, and training.

3. Implement secure communication protocols [3,8,9,22,28]

One of the biggest challenges in securing IIoT devices is ensuring the security of communication between devices and networks. To address this challenge, organizations should implement secure communication protocols such as Transport Layer Security (TLS) or Secure Sockets Layer (SSL). These protocols provide secure communication by encrypting data in transit and verifying the identity of the communicating parties.

Additionally, organizations should ensure that communication between devices and networks is encrypted end-to-end. This means that data is encrypted at the device level and remains encrypted until it reaches its destination, providing an additional layer of security.

4. Implement access controls [8–10,27,28]

Access controls are critical for securing IIoT devices. Organizations should implement access controls that limit access to IIoT devices to only authorized personnel. This can be achieved through the use of strong passwords, multi-factor authentication, and role-based access control.

In addition, organizations should monitor access to IIoT devices and networks and implement intrusion detection and prevention systems to detect and mitigate unauthorized access attempts.

5. Implement device hardening [6,8,27,28]

Device hardening involves configuring IIoT devices to minimize their attack surface. This involves disabling unnecessary services and ports, changing default passwords, and updating firmware and software regularly.

In addition, organizations should ensure that IIoT devices are configured with the minimum necessary privileges to perform their intended functions. This can be achieved through the use of access controls and role-based access control.

6. Implement security updates and patches [8,21,27,28]

Regularly updating and patching IIoT devices is critical for maintaining their security. This involves applying security updates and patches as soon as they become available.

To implement this best practice, organizations should establish a patch management process that includes regularly scanning IIoT devices for vulnerabilities, testing patches in a non-production environment, and deploying patches as soon as possible.

7. Implement security monitoring and incident response [8,27,28]

Security monitoring and incident response are critical for detecting and responding to security incidents involving IIoT devices. Organizations should implement security monitoring tools that can detect security incidents in real time and alert security personnel.

In addition, organizations should establish an incident response plan that includes procedures for containing and mitigating security incidents involving IIoT devices. This plan should be tested regularly to ensure its effectiveness.

8. Implement security awareness training [8,9,27,28,34]

As discussed earlier, human error is one of the biggest challenges in securing IIoT devices. Implementing security awareness training for all personnel who interact with IIoT devices can help mitigate this risk. This training should cover topics such as password hygiene, phishing attacks, and social engineering.

In addition, organizations should establish policies and procedures for reporting security incidents and suspicious activity. This can help ensure that security incidents are detected and addressed in a timely manner.

9. Regularly assess and audit security measures [8,27,28]

Regularly assessing and auditing security measures is critical for ensuring that IIoT devices remain secure over time. Organizations should conduct regular security assessments and penetration testing to identify vulnerabilities and assess the effectiveness of security measures.

In addition, organizations should establish a process for auditing the configuration of IIoT devices to ensure that they are configured in accordance with security policies and best practices.

The security challenges associated with Industrial IoT devices are complex and multifaceted. Organizations that fail to address these challenges risk significant financial, operational, and reputational harm. However, by implementing best practices such as a risk-based approach, defense-in-depth strategy, secure communication protocols, access controls, device hardening, security updates and patches, security monitoring and incident response, security awareness training, and regular security assessments and audits, organizations can significantly improve the security posture of their IIoT devices.

It is also important for organizations to stay up-to-date with the latest security trends and best practices and to continually reassess their security posture in light of new threats and vulnerabilities. By taking a proactive approach to security, organizations can effectively manage the risks associated with Industrial IoT devices and ensure that these devices continue to provide value to their operations.

In conclusion, Industrial IoT devices have the potential to transform industrial operations and improve efficiency, but they also present significant security challenges. As the use of these devices continues to grow, it is important for industrial organizations to take a proactive approach to security and implement strong security measures to prevent unauthorized access and quickly detect and respond to security incidents. By doing so, industrial organizations can help ensure the safety and reliability of their operations and protect against the potentially devastating consequences of cyberattacks.

10.6 FUTURE DIRECTIONS

As Industrial IoT devices continue to proliferate and integrate further into the fabric of industrial operations, the security landscape is likely to evolve in several ways. In this section, we will explore some future trends in IIoT security and what organizations can do to prepare for these changes.

1. Increased adoption of AI and machine learning

One trend likely to impact IIoT security is the increased adoption of artificial intelligence (AI) and machine learning (ML) technologies. These technologies have the potential to revolutionize industrial operations by enabling real-time data analysis, predictive maintenance, and autonomous decision-making. However, they also introduce new security risks.

AI and ML models can be vulnerable to attacks such as adversarial attacks, where an attacker manipulates data to deceive the model, or poisoning attacks, where an attacker introduces malicious data into the training dataset to compromise the model. Organizations adopting AI and ML technologies will need to secure these technologies against these and other types of attacks.

One approach is to use techniques such as data validation, model explainability, and anomaly detection to identify and address potential security issues. In addition, organizations should consider implementing secure data-sharing protocols and access controls to protect sensitive data.

2. Greater use of blockchain technology

Another trend likely to impact IIoT security is the increased adoption of blockchain technology. Blockchain is a distributed ledger technology that enables secure and transparent data sharing across multiple parties. It has the potential to revolutionize supply chain management, asset tracking, and other industrial applications.

One of the key benefits of blockchain technology is its ability to provide a tamper-evident and immutable record of transactions. This can help prevent data tampering and ensure the integrity of data across multiple parties. However, blockchain technology is not a silver bullet for IIoT security and comes with its own set of security challenges.

For example, blockchain technology can be vulnerable to attacks such as 51% attacks, where an attacker gains control of the majority of computing power on the network and can manipulate transactions. In addition, organizations adopting blockchain technology will need to ensure robust access controls and encryption protocols are in place to protect sensitive data.

3. Increased use of edge computing

Edge computing is another trend likely to impact IIoT security. Edge computing involves processing data closer to the source, rather than sending all data to a central data center for processing. This can help reduce latency and improve the responsiveness of IIoT applications.

However, edge computing also introduces new security challenges. Edge devices may be more vulnerable to physical attacks, and securing them can be more difficult than securing centralized data centers. In addition, edge devices may have limited computing resources, which can make it challenging to implement advanced security measures.

Organizations adopting edge computing will need to take a risk-based approach to security and implement a defense-in-depth strategy that includes access controls, secure communication protocols, device hardening, security updates and patches, and security monitoring and incident response.

4. Greater focus on supply chain security

Supply chain security is becoming an increasingly important issue for IIoT devices. IIoT devices typically have complex supply chains that involve multiple vendors and suppliers. This can make it challenging to ensure the security of all components of an IIoT.

In addition, the rise of counterfeit components and malicious firmware poses a significant threat to IIoT security. Attackers can exploit these components to compromise IIoT devices and gain unauthorized access to industrial systems.

To address these challenges, organizations will need to adopt a more proactive approach to supply chain security. This may include implementing secure supply chain management practices, such as vendor risk assessments and supplier security audits. Moreover, organizations may need to implement hardware and firmware authentication mechanisms to ensure that only trusted components are used in IIoT devices.

5. Standardization and regulation

As IIoT devices become more widespread, there is likely to be an increased focus on standardization and regulation. Governments and industry organizations may develop standards and regulations to ensure that IIoT devices are secure and meet certain security requirements.

For example, the US government has developed the NIST Cybersecurity Framework, which provides guidelines and best practices for managing cybersecurity risks. The European Union has also developed the General Data Protection Regulation (GDPR), which regulates the processing of personal data.

In addition, industry organizations such as the Industrial Internet Consortium (IIC) and the International Society of Automation (ISA) have developed standards and best practices for IIoT security.

Organizations that adopt IIoT devices will need to stay informed about these developments and ensure compliance with any relevant standards and regulations.

6. Increased focus on security awareness and training

As IIoT security becomes more complex, there will likely be an increased focus on security awareness and training. Organizations must ensure that their employees are aware of the risks associated with IIoT devices and are trained to identify and respond to security incidents.

This may include developing security awareness and training programs covering topics such as social engineering, phishing attacks, and device security. In addition, organizations may need to implement security policies and procedures governing the use of IIoT devices to ensure that all employees understand and adhere to these policies.

The security challenges associated with IIoT devices are complex and multifaceted. As IIoT devices continue to proliferate and integrate further into industrial operations, organizations will need to adopt a proactive and risk-based approach to security.

This approach may include implementing a defense-in-depth strategy including access controls, secure communication protocols, device hardening, security updates and patches, and robust security monitoring and incident response. In addition, organizations may need to stay abreast of emerging security trends and technologies, such as AI and machine learning, blockchain, and edge computing, and be prepared to address these challenges.

Ultimately, the success of IIoT devices in industrial applications will depend on the ability of organizations to effectively manage security risks, ensuring the integrity, confidentiality, and availability of industrial systems and data.

10.7 CONCLUSION

The security challenges associated with Industrial IoT devices are significant and multifaceted. The increasing use of IIoT devices in industrial applications has brought about new risks and threats that must be addressed through proactive and risk-based security strategies.

As discussed in this chapter, some of the key security challenges associated with IIoT devices include:

- Vulnerabilities in IIoT devices and systems
- Lack of secure communication protocols
- Weak access controls and authentication mechanisms
- Difficulty in updating and patching IIoT devices
- Standards and regulatory challenges
- Increased focus on security awareness and training
- Emerging security trends such as AI and ML

To address these challenges, organizations must adopt a defense-in-depth approach to security that includes multiple layers of security controls and measures. This may include implementing access controls, secure communication protocols, device hardening, security updates and patches, and security monitoring and incident response.

In addition, organizations must stay abreast of emerging security trends and technologies, such as AI and machine learning, blockchain, and edge computing, and ensure they are prepared to address these challenges.

Furthermore, increased collaboration between industry organizations, governments, and academic institutions is needed to develop standards and regulations that address the unique security challenges of IIoT devices. This includes developing guidelines and best practices for managing cybersecurity risks and developing technical standards for IIoT security.

Finally, organizations must prioritize security awareness and training to ensure that their employees are aware of the risks associated with IIoT devices and are equipped to identify and respond to security incidents.

As IIoT devices become increasingly prevalent in industrial settings, it is crucial for organizations to prioritize security and invest in developing secure and resilient IIoT systems. By adopting a proactive and holistic approach to IIoT security, we can fully realize the benefits of IIoT devices while minimizing the risks of security breaches and disruptions.

REFERENCES

1. Jha, S., N. Jha, D. Prashar, S. Ahmad, B. Alouffi, and A. Alharbi. (2022). "Integrated IoT-based secure and efficient key management framework using hashgraphs for autonomous vehicles to ensure road safety," *Sensors* 22(7): 2529.
2. Yu X. and H. Guo (2019). "A survey on IIoT security," *2019 IEEE VTS Asia Pacific Wireless Communications Symposium (APWCS)*. https://doi.org/10.1109/vts-apwcs.2019.8851679.
3. "Industrial Internet of Things (IIoT) – Definition," *Trendmicro.com*, 2015. https://www.trendmicro.com/vinfo/us/security/definition/industrial-internet-of-things-iiot (accessed May 13, 2023).
4. Chen, F. and Y. Luo. (2017). *Industrial IoT Technologies and Applications.* Cham: Springer International Publishing. https://doi.org/10.1007/978-3-319-60753-5.
5. Ahmad, S., S. Jha, A. Alam, M. Alharbi, and J. Nazeer. (2022). *Analysis of Intrusion Detection Approaches for Network Traffic Anomalies with Comparative Analysis on Botnets (2008–2020).* London: Security and Communication Networks.
6. Knapp, E. D. and J. T. Langill. (2014). "Hacking industrial control systems," In *Industrial Network Security*, T. Simonite, Ed. Great Britain: Syngress, pp. 171–207. https://doi.org/10.1016/b978-0-12-420114-9.00007-1.
7. Jiang, X., M. Lora, and S. Chattopadhyay. (2020). "An experimental analysis of security vulnerabilities in industrial IoT devices," *ACM Transactions on Internet Technology* 20(2): 1–24. https://doi.org/10.1145/3379542.
8. Khan, R., P. Maynard, K. McLaughlin, D. M. Laverty, and S. Sezer. (2016). "Threat analysis of blackenergy malware for synchrophasor based real-time control and monitoring in smart grid," *Electronic Workshops in Computing.* https://doi.org/10.14236/ewic/ics2016.7.
9. Adamov A. and A. Carlsson. (2017). "The state of ransomware. Trends and mitigation techniques," *2017 IEEE East-West Design & Test Symposium (EWDTS).* https://doi.org/10.1109/ewdts.2017.8110056.
10. Fayi, S. Y. A. (2018). "What Petya/NotPetya ransomware is and what its remidiations are," *Advances in Intelligent Systems and Computing* 2018: 93–100. https://doi.org/10.1007/978-3-319-77028-4_15.
11. Bajramovic, E., D. Gupta, Y. Guo, K. Waedt, and A. Bajramovic. (2020). "Security Challenges and Best Practices for IIoT," *5th International Conference on Smart and Sustainable Technologies (SpliTech), Split, Croatia, 2020.* https://doi.org/10.18420/inf2019_ws28.
12. Mosteiro-Sanchez, A., M. Barcelo, J. Astorga, and A. Urbieta. (2020). "Securing IIoT using defence-in-depth: Towards an end-to-end secure Industry 4.0," *Journal of Manufacturing Systems* 57: 367–378. https://doi.org/10.1016/j.jmsy.2020.10.011.
13. Barbareschi, M., V. Casola, A. De Benedictis, E. La Montagna, and N. Mazzocca. (2021). "On the adoption of physically unclonable functions to secure IIoT devices," *IEEE Transactions on Industrial Informatics.* 17: 1–1. https://doi.org/10.1109/tii.2021.3059656.

14. Serror, M., S. Hack, M. Henze, M. Schuba, and K. Wehrle. (2020). "Challenges and opportunities in securing the Industrial Internet of Things," *IEEE Transactions on Industrial Informatics* 17: 1–1. https://doi.org/10.1109/tii.2020.3023507.

15. Bajramovic, E., D. Gupta, Y. Guo, K. Waedt, and A. Bajramovic, "Security Challenges and Best Practices for IIoT." 243–254. ISSN:1617-5468, ISBN: 978-3-88579-689-3. https://doi.org/10.18420/inf2019_ws28.

16. Sari, A., A. Lekidis, and I. Butun. (2020). "Industrial networks and IIoT: Now and future trends," in *Industrial IoT*, I. Butun, Ed. Switzerland: Springer, pp. 3–55. https://doi.org/10.1007/978-3-030-42500-5_1.

17. Jaidka, H., N. Sharma, and R. Singh. (2020). "Evolution of IoT to IIoT: Applications & challenges," *Proceedings of the International Conference on Innovative Computing & Communications (ICICC)* 2020: 3603739. https://doi.org/10.2139/ssrn.3603739.

18. Iyyappan, M., A. Kumar, S. Ahmad, S. Jha, B. Alouffi, and A. Alharbi. (2023). "A component selection framework of cohesion and coupling metrics," *Computer Systems Science and Engineering* 44(1): 351–365.

19. Paul, S. P., S. Aggarwal, and S. Jha. (2022). "A survey on the use of artificial intelligence (AI)-enabled techniques in digital marketing," *AIP Conference Proceedings*, vol. 2555, no. 1, p. 050003. AIP Publishing LLC, New York.

20. Rajagopal, A., S. Jha, R. Alagarsamy, S. G. Quek, and G. Selvachandran. (2022). "A novel hybrid machine learning framework for the prediction of diabetes with context-customized regularization and prediction procedures," *Mathematics and Computers in Simulation* 198: 388–406.

21. Singh, K. and S. Jha. (2021). "Lean manufacturing – An analytical approach towards Industry 4.0," *2021 2nd International Conference on Smart Electronics and Communication (ICOSEC), Trichy, India*, pp. 1690–1695. https://doi.org/10.1109/ICOSEC51865.2021.9591684.

22. Anandan, R. G. Suseendran, S. Pal, and N. Zaman. (2021). *Industrial Internet of Things (IIoT): Intelligent Analytics for Predictive Maintenance.* Hoboken, NJ: Wiley-Scrivener.

23. Butun, I., M. Almgren, V. Gulisano, and M. Papatriantafilou. (2020). "Intrusion detection in industrial networks via data streaming," In *Industrial IoT*, I. Butun, Ed. Cham, Switzerland: Springer, pp. 213–238. https://doi.org/10.1007/978-3-030-42500-5_6.

24. Antonakakis, M. et al., (2017). "Understanding the Mirai Botnet Open access to the Proceedings of the 26th USENIX Security Symposium is sponsored by USENIX Understanding the Mirai Botnet,". Available: https://www.usenix.org/system/files/conference/usenixsecurity17/sec17-antonakakis.pdf.

25. Triton (2017). "International cyber law: Interactive toolkit," *International Cyber Law: Interactive Toolkit,* Jun. 04, 2021. https://cyberlaw.ccdcoe.org/wiki/Triton_(2017) (accessed May 13, 2023).

26. "TRITON malware targeting safety controllers," *Ncsc.gov.uk*, 2023. https://www.ncsc.gov.uk/information/triton-malware-targeting-safety-controllers (accessed May 13, 2023).

27. Cherepanov A. and R. Lipovsky. (2016) "Blackenergy – What we really know about the notorious cyber attacks." Available: https://www.virusbulletin.com/uploads/pdf/magazine/2016/VB2016-Cherepanov-Lipovsky.pdf

28. Ahmad, S., S. Jha, H. A. M. Abdeljaber, M. K. I. Rahmani, M. M. Waris, A. Singh, and M. Yaseen. (2022). "An integration of IoT, IoC, and IoE towards building a green society," *Scientific Programming 2022*. https://doi.org/10.1155/2022/2673753]

29. Panchal, A. C., V. M. Khadse, and P. N. Mahalle. (2018). "Security issues in IIoT: A comprehensive survey of attacks on IIoT and its countermeasures," *2018 IEEE Global Conference on Wireless Computing and Networking (GCWCN)*. https://doi.org/10.1109/gcwcn.2018.8668630.

30. "Industrial IoT security: Challenges, solutions and devices | SaM Solutions," *SaM Solutions*, Feb. 20, 2023. https://www.sam-solutions.com/blog/industrial-iot-security/ (accessed May 13, 2023).

31. "Industrial IoT: Threats and countermeasures," *Rambus*, Mar. 05, 2020. https://www.rambus.com/iot/industrial-iot/ (accessed May 13, 2023).

32. "Stuxnet | Definition, origin, attack, & facts | Britannica," *Encyclopædia Britannica*. 2023. [Online]. Available: https://www.britannica.com/technology/Stuxnet (accessed May 13, 2023).

33. Farwell, J. P. and R. Rohozinski. (2011). "Stuxnet and the future of cyber war," *Survival* 53(1): 23–40. https://doi.org/10.1080/00396338.2011.555586.

34. Shu, X., K. Tian, A. Ciambrone, and D. Yao. (2017). "Breaking the target: An analysis of target data breach and lessons learned." Available: https://arxiv.org/pdf/1701.04940.pdf.

35. Jones, C. (2021). "Warnings (& lessons) of the 2013 target data breach," *Red River | Technology Decisions Aren't Black and White. Think Red*. https://redriver.com/security/target-data-breach (accessed March 13, 2023).

36. Arya, A. and S. Jha. (2021). "An analytical review on data privacy and anonymity in "Internet of Things (IoT) enabled services". *2021 3rd International Conference on Advances in Computing, Communication Control and Networking (ICAC3N), Greater Noida, India*, pp. 1432–1438. https://doi.org/10.1109/ICAC3N53548.2021.9725770.

37. McCoy, K. (2017). "Target to pay $18.5M for 2013 data breach that affected 41 million consumers," *USA TODAY*, May 23, 2017. https://www.usatoday.com/story/money/2017/05/23/target-pay-185m-2013-data-breach-affected-consumers/102063932/ (accessed May 13, 2023).

38. Security risks and challenges to IoT devices. https://images.app.goo.gl/ZhjJGX6z4UqJq8D18.

Chapter 11

Blockchain technology, Bitcoin, and IoT

P. Srinivas Kumar, Jyotir Moy Chatterjee,
Abhishek Kumar, Pramod Rathore, and R. Sujatha

11.1 INTRODUCTION

The blockchain is the latest creation in which an individual or a group of unique personalities are treated as a pen name [1]. By transforming average data into correct data, but not the same, i.e., replicated, blockchain design has taken development to another level. Firstly, associated with Bitcoin [2], innovation specialists are discovering the solutions to improve innovation by connecting with different advancements or ideas. Bitcoin has been termed "digital gold," for giving correct and trustable results. To make the results more accurate, the United States has been spending more on developing the software. Blockchain results will be quite different from normal estimations. Unlike the internet (or a vehicle), we don't need to know the blockchain concept to utilize it efficiently. Notwithstanding, the basic knowledge of new development demonstrates how it can be seen as evolving through stages.

11.1.1 Blockchain durability and power

Blockchain itself reflects the development of web applications toward the internet by gathering all the data into the blocks. The blockchains cannot:

- Be managed by any substance.
- Have any single use for disappointment.

The concept of Bitcoin was developed in 2008. From the beginning, this project was a significant undertaking without any problems for the users. (For any queries regarding blockchain, Bitcoin was expected to be hacked or else botched. As it were, these problems started from awful things and humans made mistakes that did not affect the basic problems.) The internet came into existence nearly 30 years ago. The fame of this concept, which looks great for blockchain innovation, has developed [3], tried to provide a state-of-the-art survey of Bitcoin-related technologies and sum up various challenges.

DOI: 10.1201/9781003530572-11

11.1.2 The things connect over a secured validation procedure

Blockchain technology enables the maintenance of agreement among participants through a self-verifying process at regular intervals. This self-evaluating system of digital value ensures transactions occur rapidly. Each transaction is recorded in "blocks," forming a chain of data. Key insights and conclusions are derived from these recorded transactions. These are:

- The data that we store is public to the users; it follows the principle of transparency.
- Blockchain cannot modify any type of information; modifying over the blockchain should use extensively in figuring the overall device.

In principle, this mechanism can be flexible, but it is unlikely to happen by chance. Taking action against the architecture for catching Bitcoins, for example, would likewise have the impact of obliterating their esteem. (The device linked to the blockchain organizes and utilizes the errand of accepting along with taking charges of transfers.) The device generates a duplicate copy of the blockchain that gets loaded naturally later in the blockchain arrangement. By combining the procedure, they make the process incredible to a second-position device, which is overall the same as how the website works.

The entire system leverages blockchain technology, enabling access and storage of data from Bitcoin exchanges in decentralized databases. This mechanism is transparent to both devices and even the most skilled experts. The absence of a central authority is a fundamental principle of the system. Additionally, the data exhibits a high degree of variety, while remaining readily understandable due to its well-structured format. This allows for easy analysis and examination.

11.1.3 Who utilizes the concept of blockchain?

As a part of web designing, we do not need to think about blockchain, even though it can be beneficial for our lives. At present, it offers the most used use cases for development. Global settlements, as reported by The World Bank, saw more than $430 billion of dollars transferred in 2015. Currently, there is a high demand for blockchain developers [4]. Blockchain erases the agent for such kinds of transfers. Personal computerization became accessible to all the members with Graphical User Interface (GUI), which is known as "work area." A well-known GUI in the blockchain is alleged to be the "wallet" tasks, where normal users purchase items from Bitcoin and save them aside with distant digital forms of money. Transferring the information from one user to another will require confirmation. It is a task that is hard to envision; the wallet function is going to vary for a long time in providing a variety of procedures for the board.

11.1.4 The blockchain and improved privacy

By transferring the information over to the server, blockchain eliminates the unsafe processes that go with that information being shrouded midway. If a gadget needs to bind together the motivations behind weakness, PC (personal computer) authorities can be mishandled. The current web has security issues that are known to everyone. The general gadget relies upon a "username" design to secure our client undertakings and sources on the web. Blockchain security techniques use encryption systems. The primary concern for the utilization of the encryption norms is that it is open access and private "keys." A "public key" is a client's area over the blockchain. Bitcoins move information to the framework and get stored with an area having an actual address. The "private key" shows a mystery key that offers authorizations to their Bitcoin and other related sources. Sparing our information over the blockchain is straightforward. This is substantial, and making secure our critical sources will be required in protecting our private key by handling it and making it accessible in the wallet. By blockchain disclosure, the web draws out another degree of convenience. To date, customers can communicate with one another or with specialist co-ops – Bitcoin transactions in 2017 showed up a midpoint of around US $2 billion every day. With blockchain, new website design and development is a greater task for developing and organizing funds. Goldman Sachs believes that blockchain development carries extraordinary potential, mainly in terms of advanced mechanisms of clearing and data balancing, that can access worldwide investment funds of up to US $6 trillion every year.

11.2 DEFINING DIGITAL TRUST

Trusting a dangerous judgment among various social groups, along with other modernized worlds, choosing trust routinely comes under exhibiting uniqueness (affirmation) as well as showing assets (endorsement). Even more essential, you have to know, 'Okay say you are who you state you are?' and 'Would it be prudent for you to have the ability to do what you are endeavoring to do?' Under blockchain development, secret key generation provides stunning ownership contraption that fulfills approval requirements. Responsibility for secret keys is ownership. It similarly spares users from sharing a lot of up close and personal data that they need for transfer, leaving them exposed to software engineers. Approval isn't adequate. Endorsement – having a lump sum amount, exploring the correct trade type, etc. – needs a distributed framework as a starting point. An appropriate framework lessens the risk of consolidated corruption or disillusionment. This scattered framework ought to be centered around the trade framework's recordkeeping along with privacy. Endorsing trades is the delayed consequence of the overall framework by proceeding with all the standards whereupon it is arranged. Affirmation along with endorsement thus considered interchanges over the electronic environment without depending upon trust. Presently, businessmen in different categories have recognized the implications of making headway – unfathomable, new and mind-blowing automated associations can be done. Blockchain development was consistently depicted as the backbone for the trade structure of the web, the basis of the web of value.

Honestly, it is likely that encrypted keys along with distributed records support customers in staying and formalizing mechanized associations, making the most of imaginative capacities. All, from governments to IT firms to banking, are hoping to amass the trading layer. Validation and acceptance, required for the latest transfers, are done by the design of blockchain discovery. The concept of the blockchain is connected to the need for a reliable set of information [5–7].

In the proof of work (PoW) [8] framework, PCs should "illustrate" that they have done the "work" of troubleshooting consistent numerical issue. If a PC successfully deals with one of these issues, it becomes eligible to add blocks to the blockchain. Nonetheless, the process of embedding blocks to the blockchain, known as "mining" in the cryptographic cash world, isn't easy. According to the blockchain data site BlockExplorer, the chances of solving these issues that Bitcoin sorts out were around more than 1 out of 7 million at the period of writing. To handle troublesome mathematical issues at possibilities, PCs should run the errands that make them costly in basic proportions of impact and essentialness. Affirmation in the undertaking doesn't make attacks for developers incomprehensible, yet rather it makes it genuinely inconsequential. In case a software engineer is expected to encourage an attack on blockchain, they need to handle troublesome measurable number related issues at 1 out of 7 trillion possibilities basically like each other individual. The cost of figuring out such an attack would almost certainly surpass the points of interest.

11.3 DIFFERENCE BETWEEN BLOCKCHAIN AND BITCOIN

The main aim of blockchain is to enable computerized data to be stored as well as disseminated, which remains unaltered. The concept can be difficult to wrap our heads around without reviewing the advancement, in actuality, so let us explore how early applications of blockchain development truly perform. Blockchain development was first outlined in 1991 by

Stuart Haber and W. Scott Stornetta, who aimed to create a framework where time-frames recorded couldn't be tampered with. However, it wasn't until practically 2008 after the launch of Bitcoin in January 2009, that the blockchain, and its primary genuine software emerged. The Bitcoin convention is based on blockchain technology. In an exploration task presenting computerized money, Bitcoin's task maker Satoshi Nakamoto alluded to it as "another digital money framework which is completely shared, with no confided in the outsider." Now, everywhere across the world, there are people who have Bitcoin. According to a recent report by Cambridge Center for other amounts, approximately 5.9 million. Suppose there are 5.9 million individuals who want to use Bitcoin on staple goods. That's the place where the blockchain exists [3].

With regards to paper cash, utilization of paper money is directed along with checks by focal experts, typically a government – however, Bitcoin isn't managed by anybody. Rather, exchanges produced in Bitcoin were checked over a system of PCs. When individuals pay for other merchandise utilizing Bitcoin, PCs on Bitcoin organize power to check exchange. To perform as such, clients perform a task on the PCs in an attempt to take care of a more difficult scientific issue, known as "hash." Whenever a PC solves the issue by "hashing" a block, its algorithmic tasks have likewise confirmed the block's exchanges. The finished exchange is publicly recorded and stored as a block on the blockchain, so, all in all, it winds up unalterable. In the case of Bitcoin, and most other blockchains, PCs that effectively confirm blocks are remunerated for their work with digital money. Even though exchanges are publicly recorded [7] on the blockchain, client information isn't – or, at any rate not in full. To lead exchanges on Bitcoin, members must run a task known as "wallet." Every wallet comprises two extraordinary as well as particular encrypted keys: a public key and a secret key. In general, a public key means exchanges deposited and withdrawn. Additionally, it appears on the blockchain record as the client's computerized mark. Regardless of whether a client gets an installment in Bitcoins to their open key, they will not have the capacity to pull it back without the help of a secret partner. A client's public key [9] is an abbreviated variant of a secret key, made by a convoluted scientific calculation. In any case, because of the unpredictability of this condition, it is practically difficult to invert the procedure and produce a private key from an open key. Therefore, blockchain innovation is viewed as classified.

11.3.1 Public keys as well as secret keys ELI5

The ELI5 ("Explain it Like I'm 5") variation: More clients on a blockchain imply that the blocks can be inserted as far as possible in the chain faster. By rationale, the blockchain of information dependably is one of the most trusted by clients. The accord convention is one of blockchain technology's most prominent qualities, yet besides, takes into consideration one of its most prominent shortcomings.

Hypothetically, it is workable for a programmer to exploit the greater part of the rule which is alluded to as a 51% attack [10–11]. Here's how it would occur. Suppose there are 54.5 million PCs in the Bitcoin organization, a basic modest representation of the truth without a doubt however a simple enough number to separate. To accomplish a lion's share of the system, a programmer needs less than 2.5 million along with multiple PCs. While performing such, an aggressor or gathering of assailants could meddle with the way toward chronicling new exchanges. They could send an exchange – along with after that invert it, influencing it to appear as if they have the coin that they just spent. This helplessness, treated as double-spending [12], is what might be compared to an ideal fake and would empower clients to spend their Bitcoins twice. Such a strike is unimaginably hard to process on a blockchain of Bitcoin's scale, as it needs hackers to get command of countless PCs. When Bitcoin was newly established in 2009 by its customers termed in the bunches, it was less requesting for assailants to handle a predominant piece of device capacity over the framework. The typical characteristic of blockchain was hailed as a limitation for emerging digital monetary forms. In *Cutting Edge Gold: Bitcoin and the Inside Story of the Misfits and Millionaires Trying to Rediscover Cash*, *New York Times* essayist Nathaniel Popper describes how the social event by customers, known as "Bitfury," pooled an expansive number of ground-breaking PCs completely to get a competitive edge on the blockchain. The main goal was to extract whatever number of blocks that would be reasonable and gain Bitcoin, which at the time were valued at around $700 each.

In the first trinary month of 2014, be that as it may, Bitfury was positioned to surpass half of the blockchain system's all-out computational power. Rather than proceeding to grow within the system, the group chose self-control and promised that they should go beyond 40%. Bitfury realized that if they continued expanding their command over the system, Bitcoin's value fell as clients sold their coins in anticipation of a likelihood of 50% assault. At the end of the day, if clients lose their confidence in the blockchain organization, the data that the organization arranges becomes useless. Blockchain clients, at that point, can just expand the system's logical performance influence to a task before it starts losing cash [13].

11.4 ADVANTAGES OF BLOCKCHAIN

For all its multifaceted nature, blockchain's capacity as a peer-to-peer type of information has been nearly unbounded. With more prominent client protection and increased security, along with lower preparation expenses and fewer blunders, blockchain innovation might just observe applications past those delineated previously. Here are the key purposes of blockchain for organizations available at present.

11.4.1 Accuracy

Trades over blockchain association were avowed by an arrangement of a large number of PCs. This clears essentially all human interventions in the verification strategy, achieving less human botch and a continuously definite record of information. Whether or not a PC over framework is to submit a framework blunder, a slip-up would simply be made to one copy of the blockchain. All together for that botch to be moved to whatever is left of the blockchain, it should affect at any rate 51% of the framework's PCs – which is nearly impossible.

11.4.2 Cost

Ordinarily, customers pay a bank to check a trade, a legal official to sign a report, or a priest to perform an undertaking. Blockchain eliminates the need for a third-party affirmation and, with it, their associated costs. Business people incur a minimal charge at any point they recognize portions using MasterCard, for example, since banks need to process those trades. However, Bitcoin doesn't have a central authority and has in every practical sense, no trade charges.

11.4.3 Decentralization

Blockchain doesn't store any client information in a central location. Or maybe, the blockchain is copied and distributed over an arrangement of PCs. Whenever a new block is added to the blockchain, every PC on the framework revives the blockchain to reflect the change. By extending information over a framework, instead of taking care of it in one central dataset, blockchain ends up being continuously difficult to adjust. In case a copy of the blockchain fell heavily influenced by a software engineer, simply a lone copy of the information, instead of the entire framework, would be jeopardized.

11.4.4 Efficiency

Trades processed through a central authority can take up to a few days to settle. For example, if you consider a Friday night processing, you can't generally watch resources in your record until Monday morning. While financial establishments operate during business hours, typically five days a week, blockchain operates 24 hours a day, seven days a week.

11.4.5 Privacy

Various blockchain frameworks work like open servers, suggesting anyone with a web affiliation can give a once-over of the framework's trade history. While customers can access the details of knowledge concerning trades, they can't access the information concerning customers performing such trades. It's a normal phenomenon where blockchain frameworks like Bitcoin remain obscure, which in truth were simply a mystery. That is, when a customer makes open trades, their unique code known as the public key [9] is stored on the blockchain, rather than the client's information. Although a person's character can still be associated with their blockchain address, this shields software engineers from getting a customer's personal information, which is a risk when financial servers are hacked.

11.4.6 Security

When a trade is recorded, its authenticity ought to be affirmed by the blockchain association. A considerable number of PCs across blockchain compete to verify that the nuances of the purchase are correct. Once the PC endorses the trade, it is incorporated into the blockchain as a square. Each square on the blockchain contains its special unique hashes, close by the outstanding hash of the square before it. Whenever point information on a square is modified at any limit, that square's hashcode changes. Straightforwardness: despite the fact that singular information on the blockchain remain hidden, the advancement itself is frequently openly accessible which infers that customers on the blockchain framework can adjust the code as they see fit, insofar as they have a bigger piece of the framework's ability skipping them. Putting away data on the blockchain which is unreservedly accessible similarly makes data tampering generously more problematic. With countless PCs on the blockchain masterminded at some irregular time, for example, it is outlandish that anyone could reveal an improvement without being observed.

11.5 LIMITATIONS OF BLOCKCHAIN

When there are important potential gains to the blockchain, there were further significant challenges to the gathering. The hindrances for utilizing blockchain advancement were not just specific. The real challenges are political just as managerial, for the most part, to state nothing of an immense number of hours (read: money) of custom programming along with information base programming expected to facilitate blockchain to introduce business frameworks. Here is a bit of trouble deterring in all cases blockchain assignment.

11.5.1 Cost

Despite the way that blockchain can save customers money on trade expenses, the development is far from cost-free. The "affirmation of undertaking" framework that Bitcoin uses to favor trades, for example, eats up massive proportions of computational force. As a general rule, energy from a colossal number of PCs on the Bitcoin association is close to what Denmark consumes every year [14]. According to an ongoing investigation [15] from think-tank Elite Fixtures, the cost of mining a singular Bitcoin varies profoundly by region, from just $531 to a staggering $26,170. Taking into account typical utility costs in the United States, that figure is closer to $4,758. Notwithstanding the costs of mining Bitcoin, customers continue driving up their capacity charges to favor trades on the blockchain. That is because when diggers embed a square to the Bitcoin blockchain, they are compensated with adequate Bitcoins in making the time productive along with imperativeness invaluable. Concerning blockchains that don't use cryptographic cash, regardless, diggers ought to be paid or for the most part, helped to support trades.

11.5.2 Inefficiency

Bitcoin is an ideal logical examination for the possible inefficient parts of the blockchain. Bitcoin's "affirmation of errand" framework takes around 600 seconds to add another square to blockchain. At this rate, it is assessed that the blockchain framework can simply regulate seven transactions per second (TPS). Though extraordinarily advanced types of cash like Ethereum (20 TPS) and Bitcoin Cash (60 TPS) have improved upon Bitcoin, they are so far compelled by blockchain. Legacy marks like Visa can process up to 24,000 TPS.

11.5.3 Privacy

While Riddle&Code on the blockchain manages to shield clients from hacks and ensure privacy, it also facilitates illegal trading and activities on the blockchain network. One notable instance of blockchain being utilized for unlawful exchanges is Silk Road, an online "lessonWeb" business center that operated from February 2011 until October 2013, when it was closed down by the FBI. The site enabled clients to browse without being sought after and make unlawful buys using Bitcoins.

11.5.4 Security

Several central banks, like [16,17], have propelled examinations concerning digital currencies as indicated by a February 2015 Bank of England research report.

11.5.5 Susceptibility

More modern digital currencies using blockchain frameworks are defenseless against 51% of attacks. These attacks are unbelievably difficult to execute due to the computational force needed to gain majority control of a blockchain plan, yet NYU software designing researcher Joseph Bonneau suggested that this could change. Bonneau released a report last year [18] assessing that 51% of attacks are presumably going to increment, as developers can now basically rent computational force, instead of buying most of the apparatus.

11.6 FUTURE OF BLOCKCHAIN

First proposed as an assessment adventure in 1991, blockchain is effectively dying down into its late twenties. Like many individuals in their 20s, blockchain has seen a lot of scrutiny in the last 20 years, with associations around the world assessing what the advancement can do and where it's going in the years to come. With various sensible applications now being executed and examined, blockchain is finally getting celebrated at age 27, in no small part due to Bitcoin and advanced cash. As a popular articulation on the tongue of each money-related authority in the nation, blockchain stands to make business and government errands dynamically definite, proficient, and secure [13].

11.7 THE BLOCKCHAIN WITH VERSION 3

Indeed.com is one of the portals for searching jobs across various categories, classifying the fields and domains that use blockchain with blockchain jobs [19]. It appears that the number of blockchain employments expanded by a stunning 207% from December 2016 to December 2017. However, that is not the end of it. As per the details, this number has expanded by substantial 631% since November 2015. Blockchain enables users to enable and provide access to servers and verifies digital information. How can we initiate new business as a result?

11.7.1 Smart contacts

Disseminated records make the mechanism of coding more powerful and secure when indicated conditions are met. Ethereal, which is a freely available mechanism of blockchain venture built explicitly for ensuring credibility, had the initial chance to use blockchains on a world-transforming scale [20].

11.7.2 The sharing economy

With cooperatives like Umber, the sharing of funding is an individual success. As of now, in any position, users who need to depend on the service providers should go through intermediaries such as Umber. By using the power of distributed installments, blockchain creates a chance to provide communication among parties in a trusting manner within a client-server mechanism model. Initially, the model OpenBazaar uses blockchain to create distributed eBay. Save the file onto our device by processing items, and we can run tasks with the help of OpenBazaar sellers without any payment transfer charges. The "no tenets" mechanisms imply that single reputation is considered imperative to the connections presently available on eBay.

11.7.3 Crowdfunding

Activities such as Kickstarter and GoFundMe are performing development tasks to support the rising community society. The popularity of the locales indicates a need for an enhancement model. The blockchain concept expands both in length and width, positively affecting publicly supported investment reserves. In 2016, Ethereum-based peer-peer autonomous consultancies raised an astounding 200 billion American dollars, and members obtained "DAO tokens" which enabled them to cast vote on a brilliant lesser time contract basis. A resulting technique where task finances explained the venture was launched without appropriate ingenuity, resulting in terrible outcomes. Nonetheless, the DAO test proposes that blockchain has a chance which is possible to refer to "another overall view in terms of financial participation."

11.7.4 Governance

By making the following results, circulated storage development could convey overall information to races of survey discussion. Ethereum-dependent shrewd numbers help to evaluate the information. The task with an authoritative primary mechanism in the occurrence of blockchain implies company efficiency becomes unquestionable when overseeing digital resources and value.

11.7.5 Supply chain

The Purchaser's task was gradually to recall all cases by ensuring all the products are authentic. Disperse the things that are real. This is achieved by utilizing blockchain-based time stamping of date and locality. For example, in contrast to the data terms, UK-dependent Provenance provides inventory signals for evaluating a chance of purchasing the product from a service provider.

11.7.6 File storage

The client-server mechanism gives the clear point of reference for the peer-to-peer communication. It conveys data across the system which shields the data from hacking. Planetary File devices make the work easy for a circulated internet to perform the task. In such a manner in which a BitTorrent transfers data over the internet, IPFS disposes of requirements of the device-connected servers. That type of enhancement is not just a helpful mechanism; it's a vital step toward improving the currently over-loaded data-conveyance mechanisms on the internet.

11.7.7 Making estimations

Public support can be a motivating factor for work that demonstrates a high-value estimation process. However, there are often unaddressed biases that can distort the true value. While expectations may influence the perceived outcome, the actual value is ultimately determined by the results of tasks, which can vary over time. Blockchains are the "insight of the group" with new developments that have no uncertainty and reveal various mechanisms yet to come. The expected advert approach of Augur makes distributed contributions on the occasion of real situations. Anyone can inquire and makes a market based on an estimated output, along with gathering half of the data transfer charges for marketing items.

11.7.8 Privacy over intellectual property

This gives rise to web servers to be comprehensively greater things than the free servers. Notwithstanding, the copyrights gathered were not formed by command over the protected discovery as a result. Shrewd contracts can give privacy and digitized closeout of imaginative internet, wiping out the typical information replicating and resharing the data. Mycelia uses a blockchain mechanism for sharing the audio files over a dissemination framework. Developed by the UK artist lyricist Imogen Heap, Mycelia performs melodies specifically to groups of onlookers, just as testing and making sovereignties to lyricists along with artists – these capacities are robotized by keen persons.

11.7.9 Internet of Things (IoT)

This section discusses how wireless sensor networks contribute to the Industrial Internet of Things (IIoT). Monitoring and Maintenance: Devices are monitored for various purposes, such as tracking the temperature of a storage room. This allows for preventative maintenance of specific electrical equipment, reducing downtime and costs.

Mechanization and Data Exchange: Wireless architectures enable flexible and adjustable mechanization processes. The combination of embedded programming, devices, and software facilitates data exchange between various components and items. This collaboration fosters the development of new products and cost optimization. It is considered a significant innovation compared to traditional power grids in terms of development, trade, and overall maintenance [21]

AT&T and the Rise of IoT: Companies like AT&T are offering solutions for the current infrastructure managed by higher-level officials. This improved infrastructure will support advanced applications like IoT products. These products will leverage big data analytics and data extraction capabilities, along with large-scale mechanized architectures, to benefit executives [22].

11.7.10 Blockchain with IoT

The digital revolution has accelerated the emergence of cutting-edge technologies like the Internet of Things (IoT). IoT connects various objects, animals, and people, enabling data exchange and interaction. The number of connected devices has grown exponentially. Recent advancements in big data, data science, and machine learning have paved the way for artificial intelligence (AI). However, AI currently relies on centralized infrastructure, owned and operated by specific providers, with probabilistic calculations.

Blockchain, a revolutionary invention by Satoshi Nakamoto, offers a decentralized alternative. In this system, the database is continuously updated, but all transactions are transparent and distributed securely using deterministic calculations. Public and private keys ensure robust security [23].

The world has seen three industrial revolutions starting from the invention of steam engine succeeded by the usage of electricity. The third revolution is the usage of electronics and IT for making production automated. IoT-based devices have been produced day in and day out consistently in recent years, and as a result, huge data are being generated. The three-tier architecture, comprised of devices, edge gateway, and cloud, makes it so simple and user-friendly for interpretation. Precisely, it is the process of integrating entire things across the globe through the internet. Collecting, sending, and receiving heterogeneous data and acting on information in a coordinated way is the wonderful technology that consist in the IoT. All leading companies are concentrating more on merging blockchain with AI to gain greater insight into the business. For the huge amount of data relying on the Internet of Things (IoT) becomes inevitable for evolving fields. The usage of blockchain is seen in all domains like financial services, rental real estate, health care, credit, insurance, federal, supply chain management, music, digital identity management, job marketplace, tourism, national security, forecasting/trading, and the list goes on with many horizons. The flow begins from data generation with IoT, and the generated data is analyzed with AI and secured via blockchain in a transparent way [24].

'Data' serves as the medium that links these three demons in a common line for optimizing the technology. Recent research findings shows that convergence of blockchain with IoT and AI facilitates a multi-billion-dollar market and it is a cutting-edge technology [25].

The significance of blockchain technology in e-business use cases is due to the following aspects:

1. The faster proliferation of blockchain technology in the e-business environment
2. The deeper and extreme connectivity and integration scenarios such as Peer-to-Peer Lending and Payments, and Smart and Collaborative Integration
3. As IoT devices connect and collaborate, this results in creation of cloud storage and big data
4. Supply chain communication is being formed and sustained in the currency exchange of international remittances
5. The rising competencies in big, fast, streaming cryptocurrencies.

Blockchain offers permanent auditable digital records and thus enables the stakeholders to view the product status. Moreover, it also provides solutions for tracking the rights and royalty payments which in turn allows the customers to get benefits in various industries wherever the activities might create an impact on the premiums. The concept of blockchain technology is trending at the core of the e-business revolution, which leads to unlimited scope for research that is not limited to any specific domain [26].

11.7.11 Neighborhood microgrids

Blockchain development rises in the performance of moving over sustainable power sources developed by trusted parties' microgrids. When the position of some of the baseboards makes vitality, Ethereal-dependent shrewd users consequently distribute it comparatively. Situated in Brooklyn, Consensys (Blockchain Technology Solutions) is the principal community which is comprehensively developing for a scope in the use of Ethereal. One of the mechanisms is collaborating over Transitive Grid and making tasks with the conveyed vitality equipment, LO3. Certain type of task was presently ready for action utilizing Ethereal shrewd users to computerize observing along resharing of microgrid vitality. The supposed "keen matrix" is a beginning reason for IoT loss.

11.7.12 Identity management

There's an unequivocal need for better size for the executives over the internet. The ability to affirm personality is a lynchpin of cash-related trades that occur over the internet. For some reason, answers regarding the privacy changes that go with online marketing were faulty, in most ideal situations. Flowed records offer overhauled procedures for exhibiting your personality, close by the probability to digitize singular files. Having a sheltered identity will in like manner be basic for online participation – for instance, in the sharing economy. Not too bad a reputation, everything takes place. Which is the most serious case for coordinating trades over the internet? Making mechanized identity standards ended up being an extraordinarily marvelous process. Specific challenges aside, a comprehensive online identity course of action requires understanding between cooperating substances and the government. The necessity to investigate legitimate devices are very important because of varying locations of the devices. Electronic business on the web at present relies upon SSL confirmation for safe trades over the internet. Nikki is a beginning-stage company that attempts to perform SSL mechanisms over the blockchain. With an initial investment of $3.5 million, Nicky estimates a thing to pack off in the middle of 2017.

11.7.13 AML and KYC

Hostile to tax evasion and know your client developers have great command in the development of blockchain. Nowadays, money-related companies should perform a task seriously by multiple-step processes for every new user. KYC sales and profits should be diminished by cross-foundation client verification along with making the observations as well as investigation viability. In the beginning, Polycot has AML/KYC ordering that makes changes to the function. Such types of modifications are done and shared among the higher authorities. Another new trade design product is called Trust in Motion. Portrayed as "Integra over KYC," Tim enables customers to have the depiction of keys and records such as international ID and service bill. When the data is stored or shared by the server, the users will check whether their data is encrypted or not and whether the data is secured or not over the blockchain.

11.7.14 Information management

These days for securing their data, the users are using online portals like Facebook for no use. Later on, users are thinking regarding the value and they are moving the information their online action creates. By the time it tends to be effectively appropriated over tiny fragmentary summing, Bitcoin – in all probability enables the formation of the amount and this will be used for this kind of transfer. The MIT venture Enigma enhances and makes the procedure enlighten from a single user to a higher level. Mystery uses encrypted systems to provide single-use data extractions from hubs in regular interval of time. Dividing the data makes Enigma adaptable. The beta packing is ensured within the next half year.

11.8 TRANSACTIONS ARE BROADCAST, AND EVERY NODE IS CREATING THEIR OWN UPDATED VERSION OF EVENTS

Blockchain procedure discusses about the development in information enrollment.

Whatever, blockchain mechanisms make an improvement so powerful, is certainly not another development or it may, be a blend of explaining improvements connected recently. It's the customized arrangement of three innovations (the Web, secret key generation, and convention administering) that makes Bitcoin maker Satoshi Nakamoto to think it is valuable.

Blockchains are built from three technologies		
1. Private Key Cryptography	2. P2P Network	3. Program (the blockchain's protocol)
Cash vs. Plastic	Tree falls in a forest	Tragedy of the commons
Identity	System of Record	Platform

The outcome was the framework in advanced collaborations Digitalized connections developed by anchoring are secure – given by the rich and powerful device engineering in blockchain development only.

11.9 CONCLUSION

Blockchain innovation has many energizing possibilities; however, some genuine contemplations should be tended to before we can say it's the innovation of things to come. Keep in mind that registering power is required to check exchanges which require high power. Bitcoin is an ideal example of the dangerous acceleration in power requested from a substantial blockchain arrangement. Albeit correct measurements on the power necessities of Bitcoin are troublesome, it's normally contrasted with little nations in its present state. That is not engaging given the present worries about environmental change, the accessibility of intensity in creating nations, and the unwavering quality of intensity in created countries. In this chapter, we have tried to provide a brief overview of blockchain technology with its various advantages, limitations, and future of this blockchain technology.

REFERENCES

1. Wikipedia Contributors. (2023, November 20). *Satoshi Nakamoto.* Wikipedia. https://en.wikipedia.org/wiki/Satoshi_Nakamoto
2. Blockgeeks. (2022, October 19). *What is Blockchain Technology? A Step-by-Step Guide for Beginners.* https://blockgeeks.com/guides/what-is-blockchain-technology/
3. Chatterjee, J. M., Ghatak, S., Kumar, R., & Khari, M. (2018). BitCoin exclusively informational money: a valuable review from 2010 to 2017. *Quality & Quantity*, 52(5), 2037–2054.
4. Blockgeeks. (2020, April 25). *How to Buy Bitcoin Anywhere! [Safe, Fast and Easy].* Blockgeeks. https://blockgeeks.com/guides/how-to-buy-bitcoin/
5. Feign, A. (2022, November 22). *What Is Blockchain Technology?* https://www.coindesk.com/learn/what-is-blockchain-technology/
6. Cambridge Judge Business School. (2020, December 17). *Global Cryptocurrency – CCAF Publications – Cambridge Judge Business School.* https://www.jbs.cam.ac.uk/faculty-research/centres/alternative-finance/publications/global-cryptocurrency/#.W1xv4NgzZ-U
7. Latest BTC Blocks. (n.d.). https://www.blockchain.com/explorer/blocks/btc
8. Nevil, S. (2023, May 27). *What Is Proof of Work (POW) in Blockchain?* Investopedia. https://www.investopedia.com/terms/p/proof-work.asp
9. Frankenfield, J. (2021, June 24). *Public Key.* Investopedia. https://www.investopedia.com/terms/p/public-key.asp
10. Carter, S. M. (2017, December 20). *Man Accidentally Threw Away $127 million in Bitcoin and Officials Won't Allow a Search.* CNBC. https://www.cnbc.com/2017/12/20/man-lost-127-million-worth-of-bitcoins-and-city-wont-let-him-look.html
11. Frankenfield, J. (2023a, June 7). *51% Attack: Definition, Who Is at Risk, Example, and Cost.* Investopedia. https://www.investopedia.com/terms/1/51-attack.asp
12. Frankenfield, J. (2023b, August 16). *Understanding Double-Spending and how to prevent attacks.* Investopedia. https://www.investopedia.com/terms/d/doublespending.asp
13. Hayes, A. (2023, April 23). *Blockchain Facts: What Is It, How It Works, and How It Can Be Used.* Investopedia. https://www.investopedia.com/terms/b/blockchain.asp
14. Lee, T.B. (2017, December 6). *Bitcoin's Insane Energy Consumption, Explained.* Ars Technica. https://arstechnica.com/tech-policy/2017/12/bitcoins-insane-energy-consumption-explained/
15. Clovr.com. (n.d.). *9 Best Bitcoin Live Casinos to Play in 2023.* Clovr. https://www.elitefixtures.com/blog/post/2683/bitcoin-mining-costs-by-country/
16. Hayes, A. (2023b, October 4). *Federal Reserve System: What It Is and How It Works.* Investopedia. https://www.investopedia.com/terms/f/federalreserve-bank.asp
17. Kenton, W. (2023, July 16). *Bank of England (BOE): Role in Monetary Policy.* Investopedia. https://www.investopedia.com/terms/b/boe.asp
18. Sujatha, R., Mareeswari, V., Chatterjee, J. M., Abd Allah, A. M., & Hassanien, A. E. (2021). A Bayesian regularized neural network for analyzing bitcoin trends. *IEEE Access*, 9, 37989–38000.

19. Blockgeeks. (2023b, November 23). *Blockchain Wiki: The Many Colorful Faces of Blockchain*. Blockgeeks. https://blockgeeks.com/guides/blockchain-jobs/
20. Blockgeeks. (2023a, September 28). *What Are Smart Contracts? [Latest Research Guide]*. https://blockgeeks.com/guides/smart-contracts/
21. ConsenSys. (2017, January 12). *5 Incredible Blockchain IoT Applications*. Blockgeeks. https://blockgeeks.com/5-incredible-blockchain-iot-applications/
22. Agrawal, R., Chatterjee, J. M., Kumar, A., & Rathore, P. S. (Eds.). (2020). *Blockchain Technology and the Internet of Things: Challenges and Applications in Bitcoin and Security*. Academic Press, London.
23. Swami, M., Verma, D., & Vishwakarma, V. P. (2020, July). Blockchain and Industrial Internet of Things: Applications for Industry 4.0. In *Proceedings of International Conference on Artificial Intelligence and Applications* (pp. 279–290). Springer, Singapore.
24. Zhao, S., Li, S., & Yao, Y. (2019). Blockchain enabled Industrial Internet of Things technology. *IEEE Transactions on Computational Social Systems*, 6(6), 1442–1453.
25. Siegfried, N., Rosenthal, T., & Benlian, A. (2020). *Blockchain and the Industrial Internet of Things: A Requirement Taxonomy and Systematic Fit Analysis* (No. 117408). Darmstadt Technical University, Department of Business Administration, Economics and Law, Institute for Business Studies (BWL).
26. Wang, Q., Zhu, X., Ni, Y., Gu, L., & Zhu, H. (2020). Blockchain for the IoT and industrial IoT: A review. *Internet of Things*, 10, 100081.

Chapter 12

IoT-based Arduino controller On-Load Tap Changer distribution transformer case study

Badri Raj Lamichhane

12.1 INTRODUCTION

12.1.1 Power distribution systems and OLTC transformers

Power distribution systems are crucial for delivering electricity from high-voltage transmission networks to end-users. Distribution transformers play a pivotal role in these systems by stepping down the voltage to a suitable level for utilization. Among the distinct types of distribution transformers, On-Load Tap Changer (OLTC) transformers can regulate the output voltage by adjusting the transformer's turn ratio through a tap changer mechanism. This feature enables efficient voltage control and ensures the optimal utilization of electrical energy [1].

12.1.2 Microcontroller and Internet of Things

The advancements in microcontroller technology have revolutionized the field of electronics by providing compact, low-power computing platforms for a wide range of applications. Microcontrollers, such as the popular Arduino platform, integrate processing power, memory, and versatile input/output capabilities into a single chip. This integration empowers microcontrollers to perform control and monitor tasks effectively. The Internet of Things (IoT) refers to the network of interconnected devices embedded with sensors, software, and communication technologies that facilitate the exchange of data. By leveraging microcontrollers and IoT connectivity, innovative solutions can be developed to enhance automation, monitoring, and control in various domains, including power distribution systems [2].

12.1.3 Need for advanced control mechanisms

Traditional OLTC transformers often rely on manual or timed tap changer operations, leading to suboptimal voltage regulation and energy losses. Additionally, monitoring transformer parameters such as voltage, current,

DOI: 10.1201/9781003530572-12

temperature, and oil level manually can be cumbersome and prone to errors. These limitations necessitate the development of advanced control mechanisms that automate tap changer operations and enable continuous monitoring of transformer conditions. By employing microcontrollers and IoT technologies, efficient and optimized control mechanisms can be implemented, offering improved performance and reliability in power distribution systems.

12.1.4 IoT-based Arduino controller for OLTC Distribution Transformers

The IoT-based Arduino controller for OLTC Distribution Transformers is a significant advancement in power distribution. This intelligent system leverages the capabilities of microcontrollers, IoT connectivity, and advanced control algorithms to address the limitations of conventional tap changer mechanisms. The controller enables real-time monitoring of critical transformer parameters, automates tap changer control based on predefined algorithms and control strategies, and facilitates remote access for monitoring and control purposes. By integrating these functionalities, the system enhances the efficiency, reliability, and safety of power distribution networks.

12.1.5 Objectives and structure of the case study report

The primary objective of this case study report is to provide a comprehensive understanding of the IoT-based Arduino controller for OLTC Distribution Transformers. The report will delve into the working principle, components, system architecture, benefits, and potential applications of this innovative technology. Through a detailed analysis, the case study aims to contribute to academic knowledge, promote the adoption of advanced control mechanisms in power distribution systems, and foster further research in this domain.

12.2 WORKING PRINCIPLE

The working principle of the IoT-based Arduino controller for OLTC Distribution Transformers involves the seamless integration of various components and functionalities. This section provides an extremely detailed explanation of each step involved in the working principle.

12.2.1 Transformer monitoring

12.2.1.1 Voltage monitoring

The IoT-based Arduino controller continuously monitors the voltage of the distribution transformer. Voltage sensors, such as potential transformers, are strategically placed at appropriate locations within the transformer to measure the voltage levels. These sensors convert the analog voltage signals into digital data. The Arduino microcontroller reads and processes this data to obtain accurate, as well as real-time information about the transformer's voltage.

12.2.1.2 Current monitoring

In addition to voltage, the controller monitors the current flowing through the transformer. Current sensors, such as current transformers or shunt resistors, are placed in the primary or secondary circuit of the transformer to measure the current levels. Like voltage monitoring, the current sensors convert analog current signals into digital data. The Arduino controller reads and processes this data to determine the actual current flowing through the transformer.

12.2.1.3 Temperature monitoring

To ensure the safe operation of the transformer, temperature monitoring is essential. Temperature sensors, such as thermocouples or resistance temperature detectors (RTDs), are installed at various critical points of the transformer, including windings, oil, and cooling systems. These sensors measure the temperature at each location, converting it into digital data. The Arduino controller collects this data and uses it to assess the thermal conditions of the transformer.

12.2.1.4 Oil level monitoring

The level of insulating oil within the transformer is monitored to ensure optimal performance and safety. Oil level sensors, such as float switches or capacitive sensors, are placed in the oil reservoir of the transformer. These sensors detect the oil level and provide digital signals to the Arduino controller. The controller processes this data to determine the oil level and triggers alerts if it falls below or exceeds the recommended level.

12.2.2 Tap changer control

12.2.2.1 Sensor data acquisition

The Arduino controller continuously receives the data from the various sensors monitoring the transformer parameters. It interfaces with the sensors through appropriate interfaces, such as analog-to-digital converters or digital communication protocols, to acquire the sensor data. The acquired data is then stored in the controller's memory for further processing.

12.2.2.2 Data processing

Upon acquiring the sensor data, the Arduino controller performs real-time data processing. This processing includes filtering to remove noise and unwanted fluctuations in the data. The data is scaled to appropriate units for analysis and control purposes. The controller also validates the data to ensure its accuracy and reliability.

12.2.2.3 Algorithm execution

Based on the processed sensor data, the Arduino controller executes predefined algorithms and control strategies. These algorithms consider the desired output voltage level and the current transformer parameters to determine the appropriate tap position for the tap changer mechanism. The algorithms utilize control theory principles and mathematical calculations to optimize the voltage regulation process.

12.2.2.4 Tap changer adjustment

Once the appropriate tap position is determined, the Arduino controller generates control signals that drive the tap changer mechanism. The controller sends these signals to the tap changer motor, which adjusts the tap position accordingly. The tap changer motor may be a stepper motor or a servo motor, depending on the specific design of the transformer. The tap changer adjustment is performed smoothly and accurately to achieve the desired voltage regulation.

12.2.3 Communication

12.2.3.1 Data transmission

The IoT-based Arduino controller establishes communication channels to transmit the sensor data, control signals, and status information. The controller utilizes communication modules such as Ethernet, Wi-Fi, or cellular connectivity to establish connections with the central monitoring station or remote access points. The data transmission is typically secured using encryption protocols to ensure the integrity and confidentiality of the information.

12.2.3.2 Real-time updates

The Arduino controller continuously updates the central monitoring station or remote access points with real-time data. This enables operators to monitor the transformer's parameters and tap changer operations remotely. The real-time updates provide immediate insights into the transformer's performance and allow for timely decision-making and proactive maintenance practices.

12.2.4 Data analysis and visualization

12.2.4.1 Data analysis techniques

The collected sensor data is analyzed using various data analysis techniques. Statistical algorithms, machine learning algorithms, and control theory methodologies are employed to identify patterns, detect anomalies, and optimize tap changer operations. These techniques help understand the transformer's behavior, predict potential issues, and improv the overall efficiency of the distribution system.

12.2.4.2 Data visualization

To present the analyzed data in an intuitive and easily understandable format, the Arduino controller employs data visualization tools. Graphs, charts, and dashboards are generated to provide operators with a visual representation of the transformer's performance, voltage regulation, and other important parameters. This visualization aids in quick decision-making and facilitates efficient monitoring of the distribution transformer.

The working principle of the IoT-based Arduino controller for OLTC Distribution Transformers encompasses multiple stages, including transformer monitoring, tap changer control, communication, and data analysis. By continuously monitoring transformer parameters, executing precise tap changer adjustments, ensuring secure communication, and providing data-driven insights, the system optimizes voltage regulation, enhances operational efficiency, enables remote monitoring and control, and facilitates proactive maintenance practices.

12.3 SYSTEM ARCHITECTURE

The system architecture section provides an in-depth analysis of the IoT-based Arduino controller for OLTC Distribution Transformers, highlighting the intricate components and their interconnections.

12.3.1 Sensor network

12.3.1.1 Voltage sensors

Voltage sensors are strategically placed within the distribution transformer to measure the voltage levels. These sensors may include potential transformers (PTs) or voltage dividers. They convert the analog voltage signals into digital data that can be processed by the Arduino microcontroller. The voltage sensors are connected to the analog input pins of the Arduino.

12.3.1.2 Current sensors

Current sensors, such as current transformers (CTs) or shunt resistors, are installed in the primary or secondary circuit of the transformer to measure the current flowing through it. These sensors convert the analog current signals into digital data. The current sensors are connected to the analog input pins of the Arduino for data acquisition.

12.3.1.3 Temperature sensors

Temperature sensors are deployed at critical points of the transformer, including windings, oil, and cooling systems, to monitor temperature variations. Thermocouples or resistance temperature detectors (RTDs) are commonly used as temperature sensors. They convert the temperature readings into digital data, which is acquired by the Arduino through analog input pins or digital communication interfaces.

12.3.1.4 Oil level sensors

Oil level sensors, such as float switches or capacitive sensors, are employed to monitor the level of insulating oil within the transformer. These sensors provide digital signals indicating the oil level. The Arduino controller receives these signals through digital input pins or specific communication interfaces.

12.3.2 Arduino controller

12.3.2.1 Microcontroller

The Arduino microcontroller serves as the central processing unit of the system. It consists of a powerful microprocessor, input/output (I/O) ports, and memory resources. The microcontroller processes the sensor data, executes control algorithms, and manages communication with external devices.

12.3.2.2 Data acquisition

The Arduino controller interfaces with the sensor network to acquire data from the voltage, current, temperature, and oil level sensors. The analog data from voltage and current sensors are sampled and converted into digital form using analog-to-digital converters (ADCs). The digital data from temperature and oil level sensors can be acquired directly through digital communication interfaces.

12.3.2.3 Data processing and control

Upon data acquisition, the Arduino controller performs real-time data processing. This includes filtering to remove noise and unwanted fluctuations from the sensor data. The processed data is then analyzed using predefined control algorithms and strategies. These algorithms consider the desired output voltage level, sensor data, and system parameters to determine the appropriate tap position for the tap changer mechanism.

12.3.2.4 Tap changer control

The Arduino controller generates control signals based on the calculated tap position. These control signals are sent to the tap changer motor, which drives the tap changer mechanism. The motor can be a stepper motor or a servo motor, depending on the specific design of the transformer. The tap changer adjusts the tap position according to the received control signals, achieving the desired voltage regulation.

12.3.3 IoT connectivity

12.3.3.1 Communication modules

The IoT-based Arduino controller utilizes communication modules to establish connectivity with the central monitoring station or remote access points. Ethernet, Wi-Fi, or cellular connectivity modules are commonly employed. These modules enable data transmission between the controller and the external devices.

12.3.3.2 Communication protocols

To ensure secure and reliable data transmission, communication protocols such as MQTT (Message Queuing Telemetry Transport) or HTTP (Hypertext Transfer Protocol) are used. These protocols facilitate the establishment of bi-directional communication channels, allowing for real-time data exchange and remote control of the OLTC Distribution Transformer.

12.3.3.3 Central monitoring station

The central monitoring station receives the sensor data and control information from the Arduino controller. It serves as the user interface for system monitoring and control. Operators can access real-time data, view graphical representations of the transformer's parameters, and make necessary adjustments remotely.

12.3.4 Power supply

12.3.4.1 Power sources

The system requires a stable power supply to operate effectively. The Arduino controller and sensor network are typically powered by a dedicated power source, such as a regulated power supply or batteries. The power source must provide the required voltage levels to ensure the proper functioning of the system components.

12.3.4.2 Power management

Power management circuitry is employed to regulate and distribute power within the system. It ensures that each component receives the appropriate power supply and prevents power-related issues such as overvoltage, undervoltage, or power fluctuations.

The system architecture of the IoT-based Arduino controller for OLTC Distribution Transformers consists of a sensor network for monitoring voltage, current, temperature, and oil level. The Arduino controller serves as the central processing unit, acquiring sensor data, executing control algorithms, and managing communication. The system also includes IoT connectivity for remote monitoring and control, a central monitoring station for data visualization, and power supply and management components to ensure stable operation. The interconnection of these components forms a robust and efficient system for optimizing voltage regulation and enhancing the overall performance of OLTC Distribution Transformers.

12.4 BENEFITS AND APPLICATIONS

This section explores the advantages and potential applications of the IoT-based Arduino controller for OLTC Distribution Transformers. It highlights the key benefits and discusses how this innovative technology can be applied in various scenarios.

12.4.1 Benefits

12.4.1.1 Enhanced operational efficiency

The IoT-based Arduino controller improves the operational efficiency of OLTC Distribution Transformers. Real-time monitoring of transformer parameters allows for proactive maintenance practices, reducing the risk of failures and optimizing performance. Automated tap changer control based on predefined algorithms ensures precise and efficient voltage regulation, minimizing energy losses and improving the overall efficiency of the distribution system.

12.4.1.2 Remote monitoring and control

The integration of IoT connectivity enables remote monitoring and control capabilities. Operators can access real-time data, monitor transformer conditions, and make informed decisions from a central monitoring station or through web-based applications. This remote access reduces the need for physical site visits and allows for prompt response to any abnormal conditions, enhancing operational efficiency and reducing maintenance costs.

12.4.1.3 Data-driven decision-making

The IoT-based system provides valuable data insights through advanced data analysis techniques. By analyzing historical and real-time data, operators can identify trends, patterns, and anomalies in the transformer's performance. This data-driven approach facilitates informed decision-making regarding maintenance schedules, load management, and system upgrades, leading to improved reliability, reduced downtime, and optimized resource allocation.

12.4.1.4 Proactive maintenance practices

Continuous monitoring of transformer parameters enables the identification of potential issues at an early stage. By detecting abnormalities in parameters such as temperature, voltage, or oil level, the system can alert operators to potential faults or failures. Proactive maintenance practices can be implemented, including prompt maintenance, repairs, or replacement of components, thereby minimizing downtime, and reducing the risk of catastrophic failures.

12.4.2 Applications

12.4.2.1 Power distribution networks

The IoT-based Arduino controller finds significant applications in power distribution networks. It can be deployed in urban, rural, or industrial settings to enhance the efficiency and reliability of the distribution system. By enabling optimized voltage regulation and remote monitoring capabilities, the system contributes to better power quality, reduced losses, and overall improved performance.

12.4.2.2 Renewable energy integration

With the increasing penetration of renewable energy sources such as solar and wind, the IoT-based Arduino controller can ease their integration into the existing distribution infrastructure. By providing real-time monitoring and control of voltage levels, the system ensures seamless integration and efficient use of renewable energy resources, contributing to a greener and more sustainable power grid.

12.4.2.3 Smart grid applications

The IoT-based controller aligns with the concept of a smart grid by offering advanced monitoring and control functionalities. It can be integrated into smart grid infrastructures to enable real-time data exchange, load management, and demand response capabilities. The system plays a crucial role in optimizing grid operations, enhancing grid stability, and supporting the integration of distributed energy resources.

12.4.2.4 Industrial applications

The IoT-based Arduino controller is also applicable in industrial settings where a reliable power supply is critical. Distribution transformers in industries often experience varying load demands, making voltage regulation essential. The system's automated tap changer control and continuous monitoring capabilities ensure a stable and efficient power supply, minimizing the risk of production disruptions and improving overall industrial productivity.

The benefits and applications of the IoT-based Arduino controller for OLTC Distribution Transformers encompass various aspects, including operational efficiency, remote monitoring, data-driven decision-making, proactive maintenance practices, power distribution networks, renewable energy integration, smart grid applications, and industrial settings. These advantages make the system a valuable solution for enhancing the performance, reliability, and sustainability of power distribution systems across different sectors.

12.5 CONCLUSION

The IoT-based Arduino controller for On-Load Tap Changer (OLTC) Distribution Transformers is a highly sophisticated and intelligent system that revolutionizes the control and monitoring of distribution transformers. By integrating microcontrollers, IoT connectivity, and precise tap changer control, this system brings numerous benefits and opens a wide range of applications in the power distribution sector.

The motivation behind developing this controller stems from the need to enhance voltage regulation capabilities in OLTC transformers while enabling real-time monitoring, remote access, and data-driven decision-making. Manual tap changing processes have been proven to be time-consuming, inefficient, and prone to human errors. The IoT-based Arduino controller overcomes these limitations by automating tap changer operations, ensuring precise voltage regulation, and continuously monitoring transformer parameters.

The system architecture encompasses a sensor network for monitoring voltage, current, temperature, and oil levels, an Arduino microcontroller for data acquisition and processing, IoT connectivity for remote access and control, and a central monitoring station for data visualization and analysis. This architecture forms a robust and interconnected framework that optimizes the performance of OLTC Distribution Transformers.

The advantages of the IoT-based Arduino controller are substantial. It enhances voltage regulation accuracy, leading to improved power quality and reliability. Real-time monitoring enables proactive maintenance and reduces downtime. Remote access capabilities allow operators to manage and control transformers from a centralized location, resulting in efficient operation and reduced operational costs. Furthermore, the data analysis and visualization features provide valuable insights into transformer performance, facilitating data-driven decision-making and optimizing power distribution systems.

The applications of this controller are diverse and extend to industrial, commercial, and residential settings where distribution transformers are utilized. It finds utility in smart grids, renewable energy systems, industrial complexes, and utility substations. The flexibility and scalability of the system make it adaptable to different transformer configurations and operational requirements.

In conclusion, the IoT-based Arduino controller for OLTC Distribution Transformers offers a transformative solution that improves voltage regulation, enables real-time monitoring and remote access, and enhances the overall efficiency and reliability of power distribution systems. With its advanced features, this system paves the way for intelligent and optimized management of distribution transformers in the modern era.

To further enhance the capabilities of this controller, continuous advancements in microcontroller technology, IoT connectivity, and data analytics should be pursued. Future research and development efforts should focus on refining algorithms, expanding communication capabilities, and integrating advanced predictive maintenance techniques. By harnessing the power of emerging technologies, the IoT-based Arduino controller has the potential to revolutionize the power distribution industry, paving the way for a smarter and more sustainable energy future.

REFERENCES

1. Arora, K. (2022). *Internet of Things-Based Arduino Controlled On-Load Tap Changer Distribution Transformer* (1st ed.). Industrial Internet of Things, Imprint CRC Press, Boca Raton, FL.
2. Young, V. Z. (2023, June 23). *Information Technology Shaping the Construction Industry*. Project Cubicle. Retrieved June 16, 2023, from https://www.projectcubicle.com/7-types-of-information-technology-shaping-the-construction-industry/

Index

Note: **Bold** page numbers refer to tables and *italic* page numbers refer to figures.

For Product Safety Concerns and Information please contact our EU
representative GPSR@taylorandfrancis.com
Taylor & Francis Verlag GmbH, Kaufingerstraße 24, 80331 München, Germany

www.ingramcontent.com/pod-product-compliance
Ingram Content Group UK Ltd.
Pitfield, Milton Keynes, MK11 3LW, UK
UKHW021120180425
457613UK00005B/165